Contents

Introduction

AQA GCSE Science Core Higher Student's Book

Welcome to the AQA GCSE Science Core Student's Book for the Higher Tier. This book has been written to support you through your first Science GCSE with AQA. The book covers all the Biology, Chemistry and Physics material as well as the key 'How science works' elements of both the new specifications A and B.

Each chapter begins with a list of **learning objectives**. Don't forget to refer back to these when checking whether you have understood the material covered in a particular chapter. **Questions** appear throughout each chapter, which will help to test your knowledge and understanding of the subject, as you go along. They will also enable you to develop key skills and understand how science works.

Activities are found throughout the book. These will take you longer to complete than the questions but will show you many of the real-life applications and implications of science. At the end of each chapter a **summary** evaluates the important points and key words. You will find the summaries useful in reviewing the work you have completed and in revising for your examinations. Don't forget to use the **index** to help you find the topic you are working on.

You will find **exam questions** at the end of each chapter, to help you prepare for your exams. These include similar questions to those in the module tests for Specification A and others like the structured questions for Specification B.

If you are studying Specification A with AQA you will take six objective tests. You will need to revise:
- Biology 1a Human Biology – chapters 1 and 2
- Biology 1b Evolution and Environment – chapters 3 and 4
- Chemistry 1a Products from Rocks – chapters 5 and 6 (sections 1 to 4)
- Chemistry 1b Oils, Earth and Atmosphere – chapters 6 (sections 5 to 9), 7 and 8
- Physics 1a Energy and Electricity – chapters 9 and 10
- Physics 1b Radiation and the Universe – chapters 11 and 12

If you are studying Specification B with AQA you will take three written papers, one for Biology (chapters 1 to 4), one for Chemistry (chapters 5 to 8) and one for Physics (chapters 9 to 12).

This book is written for the **higher-tier** examination. It will push you so that you get the best possible result at GCSE. If you are planning to take the **foundation-tier** examination, you may wish to use the Foundation Student's Book instead.

Good luck with your studies!

Nigel Heslop, Graham Hill, Toby Houghton, Steve Witney and Christine Woodward

AQA
GCSE Science
CORE HIGHER

Editor: Graham Hill

Nigel Heslop, Graham Hill,
Toby Houghton, Steve Witney
and Christine Woodward

Hodder Murray
A MEMBER OF THE HODDER HEADLINE GROUP

The Publishers would like to thank the following for permission to reproduce copyright material:

p.9 Science Photo Library/Saturn Stills; **p.10** Corbis/Sean Aidan/Eye Ubiquitous; **p.11** t Science Photo Library/AJ Photo, c Still Pictures/Michael J. Balick, b Corbis/Eric and David Hosking; **p.12** Rex Features/Mike Webster; **p.14** Bob Battersby; **p.18** Anthony Blake Photo Library/Maximilian Stock Ltd.; **p.22** l Anthony Blake Photo Library/Joy Skipper, c Anthony Blake Photo Library/Maximilian Stock Ltd, r Bob Battersby; **p.23** Science Photo Library/John Radcliffe Hospital; **p.24** l Rex Features/Henryk T. Kaiser, c Rex Features/John Powell, r Rex Features/Image Source; **p.25** l Corbis/Norbert Schaefer, c Rex Features/Image Source, r Alamy/Alaska Stock LLC; **p.30** Rex Features/TS/Keystone USA; **p.32** tl Corbis/Mediscan, tc Corbis/Visuals Unlimited, bl Science Photo Library; **p.35** Rex Features/Garo/Phanie; **p.38** Science Photo Library/Nancy Sefton; **p.39** c Rex Features/John W. Warden, b Photolibrary/OSF/David Cayless; **p.40** tl Photolibrary/OSF/Stephen Shepherd, cl Science Photo Library/John Beatty, cr Bruce Coleman Ltd, bl Ardea/Bob Gibbons; **p.41** l Rex Features/Reso, r Alamy/John Sylvester; **p.42** Alamy/Brian Elliott; **p.43** Photodisc; **p.44** Still Pictures/Mark Edwards; **p.45** l Photolibrary/OSF/Harold Taylor, cl Photolibrary/OSF/Paulo De Oliveira, cr Still Pictures/Luc Vausort, r Science Photo Library/David Aubrey, bl Alamy/Worldwide Picture Library; **p.50** t Getty Images/David Woodfall, bl Christine Woodward, br Christine Woodward; **p.51** Getty Images/AFP; **p.52** l Getty Images/Art Wolfe; **p.54** Christine Woodward; **p.56** t Christine Woodward, b Alamy/Jim West; **p.60** tl Science Photo Library/CAMR/A.B.Dowsett, tr Corbis/Stuart Westmorland, bl Rex Features/Richard Austin, br Rex Features/Rex Interstock; **p.61** Science Photo Library/James King Holmes; **p.64** t Science Photo Library/Kenneth W. Fink, b Science Photo Library/SCIMAT; **p.65** Corbis/Ian Harwood/Ecoscene; **p.67** Science Photo Library/Michael Donne; **p.68** Still Pictures/Philippe Hays; **p.69** tl Still Pictures/Martin Harvey, tr Science Photo Library/Gregory Dimijian, c Photolibrary/OSF/Sue Scott, br Mary Evans Picture Library; **p.70** Corbis/Leonard de Selva; **p.71** Hulton Archive/Getty Images; **p.72** t Corbis/Reuters, c Corbis/Reuters; **p.73** Science Photo Library/George Bernard; **p.76** Stephanie Maze/CORBIS; **p.77** t Rex Features/Nils Jorgensen, b Corbis/Chris Bland/Eye Ubiquitous; **p.80** Geoscience Features Picture Library; **p.85** Rex Features/Dave Penman; **p.87** t Corbis/Tim McGuire, b Rex Features/Sunset; **p.88** tl Russ Merne /Alamy, cl Danita Delimont/Alamy, cr Greenshoots Communications/ Alamy; **p.90** cl Philadelphia Museum of Art/CORBIS, c Photolibrary.com, cr Rob Melnychuk/Taxi/Getty Images; **p.92** Geoscience Features Picture Library; **p.93** Steve Atkins/Alamy; **p.96** Peter Bowater/Alamy; **p.97** t Pascal Goetgheluck/Science Photo Library, b Scott Camazine/Alamy; **p.103** Getty Images/Hulton Archive; **p.106** all Nigel Heslop; **p.108** both Nigel Heslop; **p.109** Still Pictures/Klaus D. Francke; **p.112** Still Pictures/Ray Pfortner; **p.114** Corbis/Peter Turnley; **p.115** cl Nigel Heslop, cr Nigel Heslop, b Bob Battersby; **p.116** all Nigel Heslop; **p.117** t Nigel Heslop, b Geoscience Features; **p.118** Nigel Heslop; **p.119** Bob Battersby; **p.120** c Photodisc, bl Nigel Heslop, bc Nigel Heslop, br Nigel Heslop; **p.121** Nigel Heslop; **p.124** t Still Pictures/Wolfgang Maria Weber, b Getty Images/UHB Trust; **p.125** t Science Photo Library/Jerry Mason, c Nigel Heslop, b Bob Battersby; **p.128** Ingram; **p.129** Ingram; **p.130** tc Photodisc, tr Rex Features/Burger/Phanie, cl Bob Battersby; **p.131** t Anthony Blake Photo Library/Tim Hill, b Bob Battersby; **p.132** Alamy/Marie-Louise Avery; **p.133** t Andrew Lambert, c Science Photo Library/Maximilian Stock Ltd., b Rex Features/The Travel Library; **p.136** Alamy/AGStockUSA, Inc; **p.140** Science Photo Library/Planetary Visions Ltd.; **p.141** cl Corbis/Jim Craigmyle, cr Getty Images/Wayne Levin; **p.149** Rex Features/RYB; **p.150** GeoScience Features; **p.151** c Getty Images/AFP, b Rex Features/Sipa Press; **p.154** Rex Features/Sipa Press; **p.155** Rex Features/The Travel Library; **p.159** cl Ecoscene/Bruce Harber, cr Courtesy Sony, bl Rex Features/Sipa Press, br Still Pictures/Hartmut Schwarzbach; **p.161** tl Science Photo Library/Gusto, bl ©Ted Kinsman / Photo Researchers, Inc. Science Photo Library; **p.165** c Andrew Ward, b Rex Features/David Cole; **p.167** Courtesy Yamaha; **p.171** c Corbis/Dimitri Iundt, b Rex Features/Phil Ball; **p.173** Lorna Ainger; **p.178** tl Corbis/Paul A. Souders, tr Still Pictures/David Woodfall/WWI, cl Science Photo Library/Tony Craddock, cr Science Photo Library/Martin Bond, bl Rex Features/Reso, br Still Pictures/Jim Wark; **p.179** Getty Images/Spencer Rowell; **p.180** t Bob Battersby, c Andrew Lambert, bl Getty Images/Luc Hautecoeur, br Getty Images/Barry Rosenthal; **p.182** Rex Features/BYB; **p.185** Science Photo Library/Simon Fraser; **p.186** Still Pictures/D.Rodrigues/UNEP; **p.188** Science Photo Library/Martin Bond; **p.189** cl Courtesy Marine Current Turbines Ltd. www.marineturbines.com, cr Ocean Power Delivery Ltd. www.oceanpd.com, b Corbis/Richard Cummins; **p.190** t Corbis/Paul Almasy, c Science Photo Library; **p.192** Courtesy LVM/Aerogen Wind Generators; **p.193** cl The Daily Telegraph 2005/Jay Williams, cr Courtesy NPower, b Still Pictures/Jorgen Schytte; **p.194** c courtesy of Better Energy Systems Ltd./Solio™, b Rex Features/Reso; **p.195** Rex Features/The Travel Library; **p.198** l Science Photo Library/Martin Dohrn, tc Still Pictures/Jorgen Schytte, bc Rex Features/Tess Peni, tr Science Photo Library/Zephyr, br Rex Features/Roy Garner; **p.199** c Empics/Steve Mitchell; **p.203** Getty Images/Katsumi Suzuki; **p.204** t Rex Features/Eye Ubiquitous, c Science Photo Library/Klaus Guldbrandsen, b Science Photo Library/Andrew Lambert Photography; **p.205** Science Photo Library; **p.206** Science Photo Library/Larry Mulvehill; **p.208** Andrew Ward; **p.211** t Science Photo Library, b Science Photo Library; **p.212** Rex Features/Image Source; **p.214** t Corbis/Underwood & Underwood; **p.216** Getty Images/AFP; **p.218** Science Photo Library/Will McIntyre; **p.219** Science Photo Library/ISM; **p.223** NASA; **p.227** Science Photo Library/Simon Fraser; **p.229** Corbis/Joseph Sohm/Visions of America; **p.231** t Science Photo Library/Tony & Daphne Hallas, b Science Photo Library/Robert Gendler.

b = bottom, c = centre, l = left, r = right, t = top

Acknowledgements
Waste Online for permission to use the pie chart data on page 123. Data in Table 6.4, page 114 reproduced from National Statistics website: www.statistics.gov.uk Crown copyright material is reproduced with the permission of the controller of HMSO.

Every effort has been made to trace all copyright holders, but if any have been inadvertently overlooked the Publishers will be pleased to make the necessary arrangements at the first opportunity.

Although every effort has been made to ensure that website addresses are correct at time of going to press, Hodder Murray cannot be held responsible for the content of any website mentioned in this book. It is sometimes possible to find a relocated web page by typing in the address of the home page for a website in the URL window of your browser.

Risk Assessment
As a service to users, a risk assessment for this text has been carried out by CLEAPSS and is available on request to the publishers. However, the publishers accept no legal responsibility on any issue arising from this risk assessment: whilst every effort has been made to check the instructions of practical work in this book, it is still the duty and legal obligation of schools to carry out their own risk assessments.

Hodder Headline's policy is to use papers that are natural, renewable and recyclable products and made from wood grown in sustainable forests. The logging and manufacturing processes are expected to conform to the environmental regulations of the country of origin.

Orders: please contact Bookpoint Ltd, 130 Milton Park, Abingdon, Oxon OX14 4SB. Telephone: (44) 01235 827720. Fax: (44) 01235 400454. Lines are open 9.00–5.00, Monday to Saturday, with a 24-hour message answering service. Visit our website at www.hoddereducation.co.uk.

© Nigel Heslop, Graham Hill, Toby Houghton, Steve Witney, Christine Woodward 2006
First published in 2006 by
Hodder Murray, an imprint of Hodder Education,
a member of the Hodder Headline Group
an Hachette Livre UK Company
338 Euston Road
London NW1 3BH

Impression number   10 9 8 7 6 5 4 3
Year   2011 2010 2009 2008 2007

Cover photos Science Photo Library: dragonfly, Andy Harmer; house, Ted Kinsman; limestone, Alfred Pasieka.
Typeset in Times 11.5pt by Fakenham Photosetting Limited, Fakenham, Norfolk
Printed in Italy

A catalogue record for this title is available from the British Library

ISBN: 978 0 340 90709 2

Chapter 1
How do our bodies respond to change?

At the end of this chapter you should:

- ✓ know how the different parts of the nervous system work together to co-ordinate our response to external stimuli;
- ✓ be able to explain the difference between a voluntary and a reflex response;
- ✓ understand that many processes in the body are co-ordinated by chemicals called hormones;
- ✓ be able to explain how hormones control the amount of water and sugar in the blood;
- ✓ be able to describe how hormones control the menstrual cycle;

- ✓ know how hormones can be used to control a woman's fertility;
- ✓ be able to explain how medical drugs are developed and tested;
- ✓ be able to describe the effects of a range of drugs on the body;
- ✓ be able to evaluate the claims of researchers about the potential risks of cannabis and its links with addiction to hard drugs.

Figure 1.1 In each of these pictures people's bodies will respond to what is happening

How do different parts of our nervous system work together and respond to change?

Look at the pictures in Figure 1.1.

1 Describe how you think each person will respond to the situations shown in the pictures.

2 For each different scenario, explain how you think the body controls its response.

Receptors are cells that can detect changes inside or outside the body.

The central nervous system (CNS) is the spinal cord and the brain.

Effectors are organs in the body that cause a response. They can be muscles or glands.

3 List four changes, either internal or external, that your body can respond to.

4 For each of the changes you have listed, write down where the receptors to detect this change are found and what the response might be. You could put your answers into a table.

Sense organs are organs that contain receptor cells.

Stimuli (singular: stimulus) are changes that receptor cells detect.

Your body is constantly responding to changes taking place outside and inside your body. If someone throws you a ball you respond quickly by moving your hands to catch it. If a teacher is about to ask you a question in class you may feel nervous. If you haven't drunk enough water on a hot day, your body will make changes to save the water you already have in your blood.

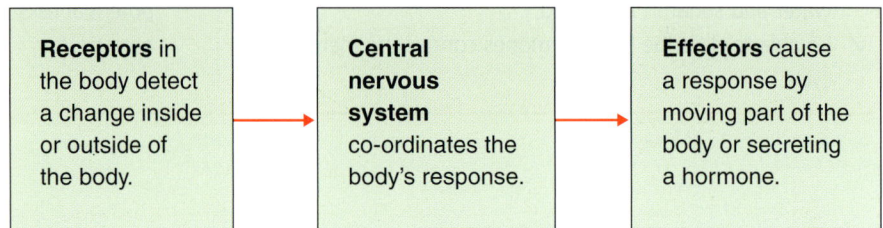

Receptors in the body detect a change inside or outside of the body.	→	Central nervous system co-ordinates the body's response.	→	Effectors cause a response by moving part of the body or secreting a hormone.

Figure 1.2 A flow diagram showing how the nervous system enables the body to respond to changes

For example, if a friend calls your name, **receptors** in your ears detect the sound. They then send a message to your **central nervous system (CNS)**. Your CNS registers that you need to turn round and so sends a message to your muscles. The muscles (the **effectors**) then contract so that you are facing your friend.

In your body, receptors are found in the **sense organs**. Eyes are the sense organs that contain light receptors. The changes that receptors detect are called **stimuli**. For example, light is the stimulus detected by the light receptors in your eyes. When a receptor detects a stimulus, it converts it into a tiny electrical impulse.

There are eight different types of receptors in your body. These are shown in Table 1.1.

Stimuli	Receptor	Sense organ
Light	Light receptors	Eye
Sound	Sound receptors	Ear
Movement	Position receptors Touch receptors Pressure receptors	Ear Skin Skin
Chemicals	Chemical receptors	Tongue Nose Blood vessels
Change in temperature	Temperature receptors	Skin Blood vessels
Pressure/temperature	Pain receptors	Skin

Table 1.1 The range of receptors found in the body and the stimuli they detect

You should now be able to explain how the nervous system responds to a range of stimuli. A good way to do this is to draw a flow diagram describing each stage. Figure 1.3 shows how the nervous system responds when you touch a hot object, like a pan on a stove.

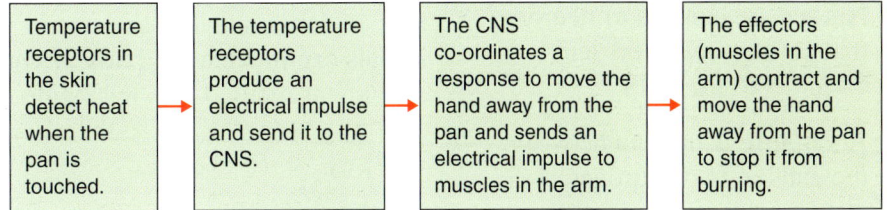

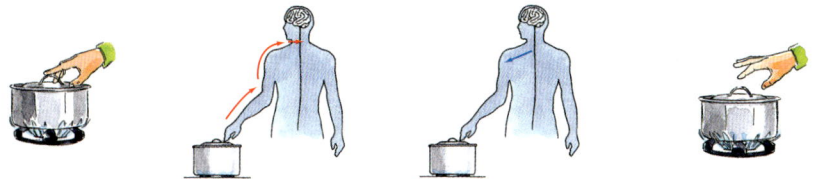

| Temperature receptors in the skin detect heat when the pan is touched. | The temperature receptors produce an electrical impulse and send it to the CNS. | The CNS co-ordinates a response to move the hand away from the pan and sends an electrical impulse to muscles in the arm. | The effectors (muscles in the arm) contract and move the hand away from the pan to stop it from burning. |

Figure 1.3 A flow diagram showing how the nervous system responds so that a person will not burn themself seriously when they touch a hot object

5 Draw flow diagrams showing how the nervous system co-ordinates a response to each situation in the three pictures in Figure 1.4. Try to include as much detail as that given in the example in Figure 1.3.

How do electrical impulses get round your body?

If you call friends who live two miles away, the phone lines connect you with your friends and carry your message to their phone. Your body connects the receptors, CNS and effectors together in a similar way. In your body, the **nerves** act like the wires in the phone lines. They conduct the electrical impulses from one place to another.

Nerves are made up of individual nerve cells called **neurones**. Neurones are laid end to end to make long strands, a bit like individual wires. These long strands are bundled together to make nerves. An electrical impulse starts at one end of a neurone and passes along it to the other end.

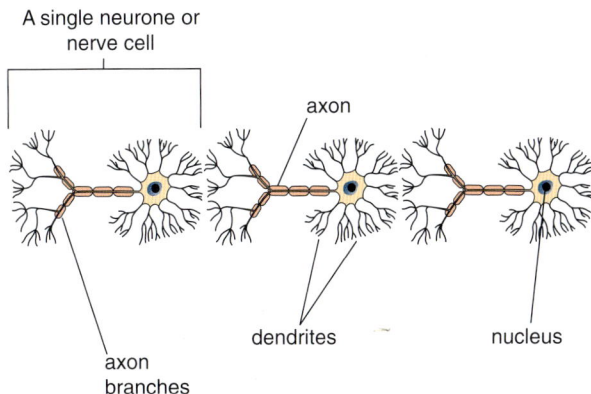

Figure 1.4 The people in each of these pictures are about to respond to stimuli they can detect

A single neurone or nerve cell

axon

axon branches

dendrites

nucleus

Figure 1.5 Three neurones forming a single strand in a nerve

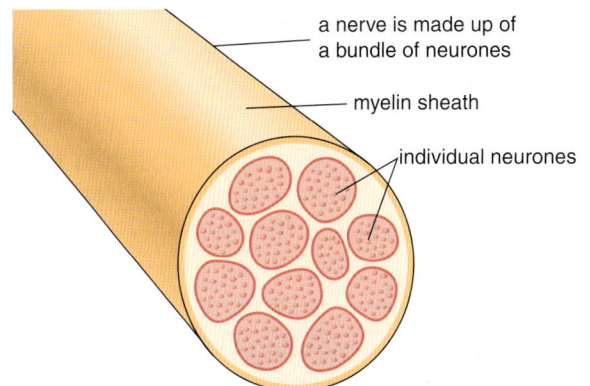

a nerve is made up of a bundle of neurones

myelin sheath

individual neurones

Figure 1.6 A bundle of neurones forms a nerve

Neurones are individual nerve cells that are specialised to transmit electrical impulses.

Nerves are bundles of neurones that connect receptors and effectors to the CNS.

A synapse is the gap between the ends of two neurones.

Chemicals that pass across a synapse and cause an electrical impulse to be generated in the next neurone are called chemical transmitters.

⑥ How is a neurone different to other types of animal cells?

⑦ How do these differences ensure a neurone is effective at transmitting electrical (nerve) impulses around the body?

⑧ Describe how information is sent from a temperature receptor in your finger to your brain.

Voluntary actions are responses that are co-ordinated by the brain.

Reflex actions are automatic, rapid responses, often to harmful situations.

Sensory neurones carry impulses from a receptor to the spinal cord.

Relay neurones carry impulses from sensory neurones to motor neurones.

Motor neurones carry impulses to effectors.

Between the end of one neurone and the start of another, there is a gap called a synapse. The electrical impulse cannot travel across this gap, but it causes the formation of a chemical transmitter which triggers an electrical impulse in the next neurone. Once the new impulse has been sent, the chemical is broken down by an enzyme. This process can happen between every neurone in your body and this ensures that the signals can move to all parts of your body.

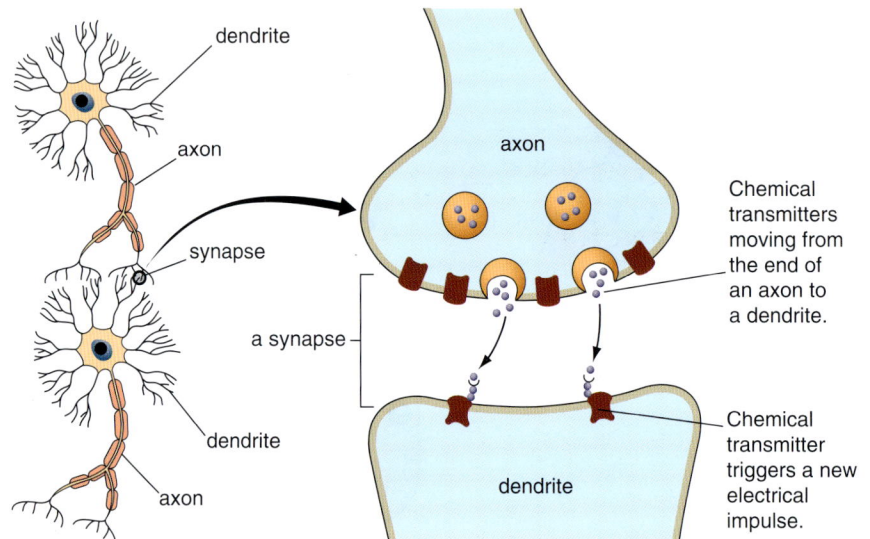

Figure 1.7 This diagram shows how a chemical transmitter is released from one neurone. This chemical travels across the synapse and then triggers the release of a new electrical impulse in the next neurone.

How fast can the nervous system respond to stimuli?

Electrical impulses can travel along your neurones at about 100 metres per second. That's the speed you would need to run to finish a 100 m race in 1 second! So, all responses are fast but for many responses you need to add the time it takes for the brain to process the information and co-ordinate a response. These responses that are co-ordinated by the brain are called voluntary actions.

Voluntary actions are fine if you are responding to a question from your friend, but if you need to respond to a harmful situation you need to respond more rapidly. For example, if you touch a sharp object, you pull your hand away before thinking about it. This type of automatic response is called a reflex action.

Reflex actions do not rely on the brain to co-ordinate a response. This is brought about by the spinal cord. A sensory neurone carries the impulse to the spinal cord where a relay neurone passes the impulse straight to a motor neurone. The motor neurone then carries the impulse to the effector to trigger a response.

9 Which of the responses listed below are voluntary actions and which are reflex actions?
- Blinking when someone kicks dust at you.
- Sneezing when pepper goes up your nose.
- Picking up a pen.
- Pulling your hand away when you touch a hot iron.
- Turning around when someone calls your name.
- Taking a CD off a shelf.

10 Choose one of the reflex actions from question 9. Draw a flow diagram similar to Figure 1.3 to show how the body co-ordinates this response.

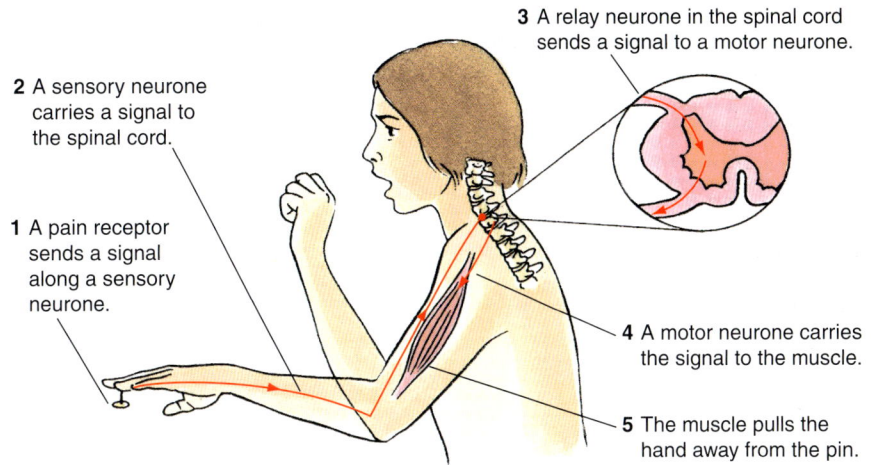

2 A sensory neurone carries a signal to the spinal cord.

1 A pain receptor sends a signal along a sensory neurone.

3 A relay neurone in the spinal cord sends a signal to a motor neurone.

4 A motor neurone carries the signal to the muscle.

5 The muscle pulls the hand away from the pin.

Figure 1.8 A reflex action brings about a quick response when someone touches a sharp object

Activity – Multiple sclerosis

Multiple sclerosis (MS) is a disease that damages the central nervous system. Its symptoms are very varied and can include: blindness, slurred speech, poor co-ordination, paralysis and memory loss. At present there is no cure for multiple sclerosis, but there is a lot of research being carried out to develop treatments for the disease.

The nerves in the central nervous system are coated with a chemical called myelin. This insulates the nerve in a similar way to the plastic coating on electrical wire. When a person suffers from MS, patches of myelin become damaged, exposing the nerve itself.

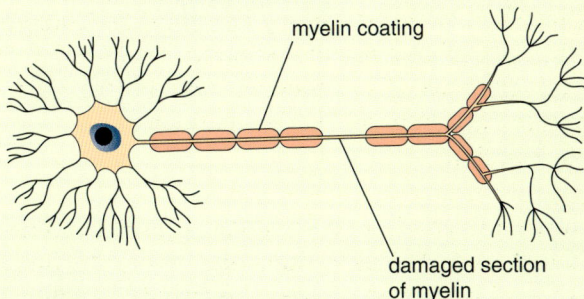

myelin coating

damaged section of myelin

Figure 1.9 A nerve with a damaged myelin coating in a person suffering from MS

1 What are the symptoms of MS?

2 From what you have learnt about the nervous system, suggest how a damaged myelin coating can cause the symptoms of MS.

3 MS only causes damage to nerves in the brain and spinal cord. However a sufferer may have poor co-ordination of their legs. Why does this happen if the nerves in the legs are not actually damaged?

4 Prepare a list of six 'frequently asked questions', with answers, that could be published on a website for people who have recently been diagnosed with MS. Think about the information they would really want to know. Try using the following websites to help you with this question:
www.mstrust.org.uk   www.mult-sclerosis.org

5 Scientists are constantly carrying out research. The increase in our scientific knowledge has helped us to answer many questions. But science cannot supply all the answers. Look carefully at the six 'frequently asked questions' that you have listed in question 4.
a) Which of these questions do you think can be answered using our present scientific knowledge?
b) Which of these questions are outside the boundaries of science? (Perhaps they are questions where beliefs, opinions or personal views are important.)

How do hormones control conditions inside your body?

Hormones are chemicals which are released from glands into the blood and then transported to their target organs.

Glands are the parts of the body that release (secrete) hormones.

A target organ is the organ that a specific hormone acts upon.

When asked about **hormones**, people often mention teenage mood swings and puberty. Both of these changes are triggered by hormones. This topic will help you to explain how hormones control lots of other processes in your body. If something scares you, or puts you under stress, your heart beats faster. This happens because a chemical called adrenaline is released into your bloodstream and then travels to your brain. Your brain responds by sending a nerve impulse to your heart, causing it to beat faster. Adrenaline is one of the many hormones produced by your body. Organs that release hormones are called **glands**. They act on specific organs called **target organs**.

How does a hormone control the water content of the body?

It is essential to have the right amount of water in your body. You take in water when you eat and drink but water is lost from your body in a number of ways. Water is lost when you breathe out, when you sweat and water leaves your body in urine when you go to the toilet.

One way in which your body can control the loss of water is by altering the amount of water in your urine. If you have too little water in the blood, your kidneys will reduce the volume of water used to make urine.

❶ Look at Figure 1.10. Name the gland, hormone and target organ in this response.

❷ Compare the response system in question 11 with the way goods are made, transported and delivered in everyday life.

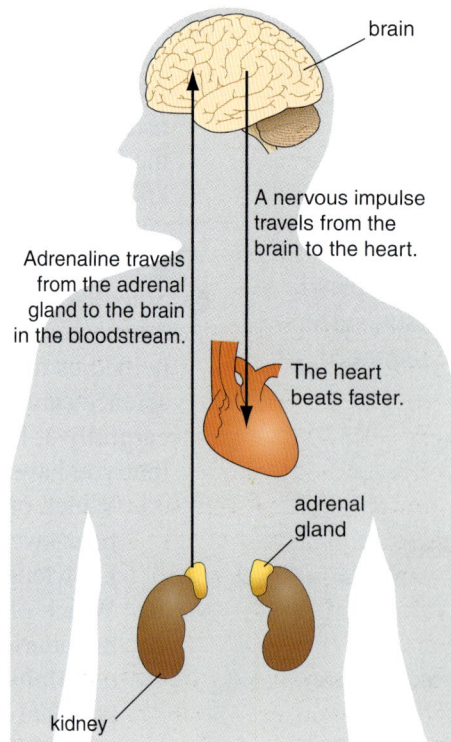

brain

A nervous impulse travels from the brain to the heart.

Adrenaline travels from the adrenal gland to the brain in the bloodstream.

The heart beats faster.

adrenal gland

kidney

Figure 1.10 How the release of adrenaline can affect the heart

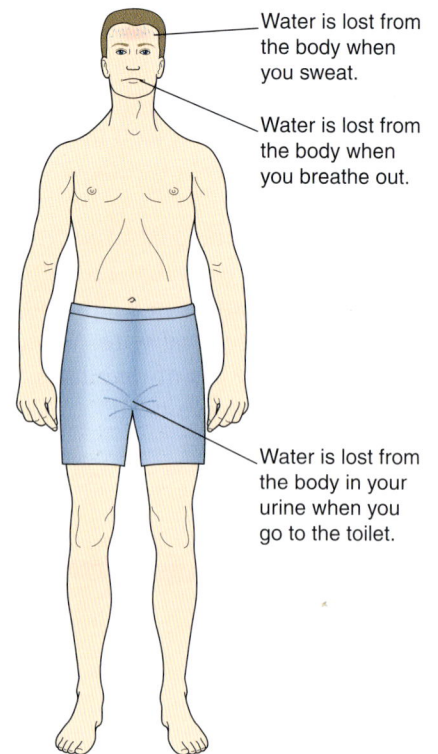

Water is lost from the body when you sweat.

Water is lost from the body when you breathe out.

Water is lost from the body in your urine when you go to the toilet.

Figure 1.11 Ways in which water is lost from the body

The water that is saved goes back into the blood. If you have too much water in the blood, your kidneys allow the excess to be lost in urine. This response in the kidneys is controlled by anti-diuretic hormone (ADH), which is secreted by the pituitary gland. ADH is carried by the blood to the kidneys, its target organs. An increase in ADH causes the kidneys to re-absorb more water into the blood. A decrease in ADH causes the kidneys to leave more water in the urine.

When water is lost from the body, in sweat and urine, essential ions such as sodium (Na^+) and potassium (K^+) are also lost. So when your body controls the amount of water in the blood it is also controlling the ion content of your body.

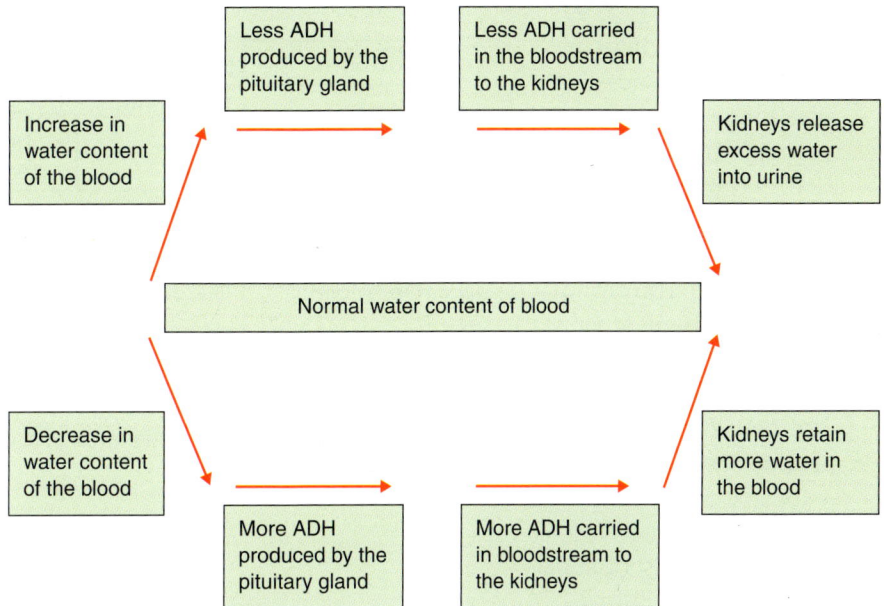

Figure 1.12 A flow chart summarising the control of water content in the blood by ADH and the kidneys

How do two hormones work together to control the concentration of sugar in your blood?

It is essential that the concentration of sugar in your blood is controlled. It should stay between 90 and 100 mg of sugar in every 100 cm³ of blood. When you eat, the amount of sugar in your blood increases. Some of this sugar is stored in the liver. It can then be released back into the blood when the blood sugar level drops. The amount of sugar that is stored is controlled by a hormone called **insulin**. The amount of sugar that is released from the liver is controlled by a hormone called **glucagon**. Both of these hormones are released from the pancreas.

Insulin is released from the pancreas when blood sugar concentration rises. This causes the liver to remove sugar from the blood and store it. When blood sugar drops too low, glucagon is released by the pancreas. This causes the liver to release stored sugar back into the blood.

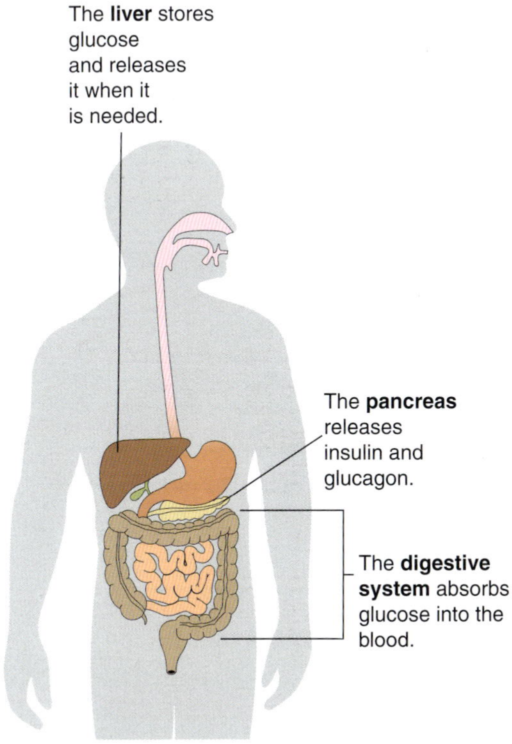

The **liver** stores glucose and releases it when it is needed.

The **pancreas** releases insulin and glucagon.

The **digestive system** absorbs glucose into the blood.

Figure 1.13 The pancreas where insulin and glucagon are released and the liver where sugar is stored

16 Name a) the gland; b) the target organ; c) the hormones involved in the control of blood sugar concentration.

17 A patient's blood sugar concentration was measured six times during a five hour period. The following values were obtained in mg of sugar per 100 cm³ of blood: 92, 89, 95, 97, 115, 97. Copy and complete the following sentences.
The range of data for the patient's blood sugar concentration varies from _____ (the lowest value) to _____ mg of sugar per 100 cm³ of blood (the highest value). One of the values appears to be anomalous. It is very different from the others and does not lie within the expected range. The anomalous value is _____ mg of sugar per 100 cm³ of blood. Ignoring the anomalous result, the patient's mean (average) blood sugar concentration was _____ mg sugar per 100 cm³ of blood.

How do hormones control the menstrual cycle in women?

A woman's menstrual cycle lasts about 28 days. During the cycle the lining of the womb (uterus) thickens. An egg matures and is released from the ovaries. If this egg is fertilised by a sperm cell it may implant in the lining of the womb. If the egg does not implant in the womb much of the lining is then shed during the woman's period. All these changes are controlled by hormones released by the ovaries and the pituitary gland in the brain (Table 1.2). The levels of these hormones in the bloodstream change throughout the menstrual cycle.

Hormone	Released by	Effect on the body
Folicle stimulating hormone (FSH)	The pituitary gland	• Causes an egg to mature in an ovary • Stimulates the ovaries to produce oestrogen
Luteinising hormone (LH)	The pituitary gland	• Stimulates the release of an egg from an ovary
Oestrogen	The ovaries	• Inhibits further production of FSH • Stimulates the pituitary gland to release LH • Causes the lining of the womb to repair and thicken

Table 1.2 The main hormones involved in control of the menstrual cycle

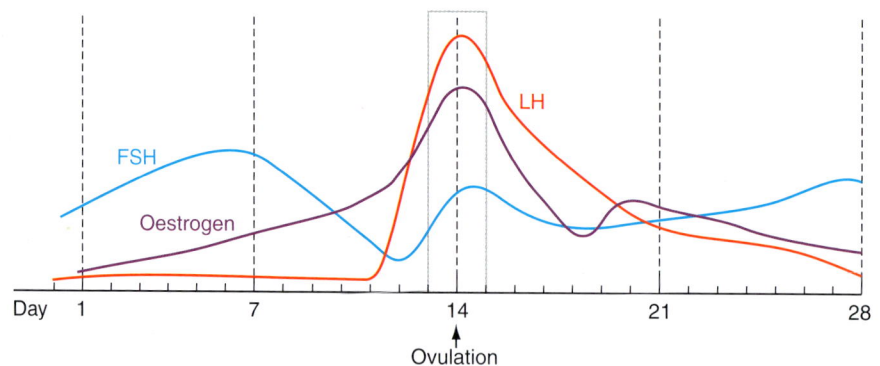

Figure 1.14 A graph showing how the levels of hormones change during the menstrual cycle

18 Sketch a copy of the graph in Figure 1.14. Add labels to it to describe the changes caused by each hormone.

Activity: Using hormones to control fertility

The hormones involved in the menstrual cycle can be used to help a woman become pregnant or to prevent this happening (contraception). Hormones in the contraceptive pill reduce fertility and prevent a woman from becoming pregnant.

Laura and David have just got married. They plan to have children in about three years' time but until then they want to have sex without getting pregnant. Laura talked to her doctor about possible methods of contraception. She suggested that Laura should go on the pill and explained that the pill contained a hormone that would prevent an egg maturing in the ovaries each month.

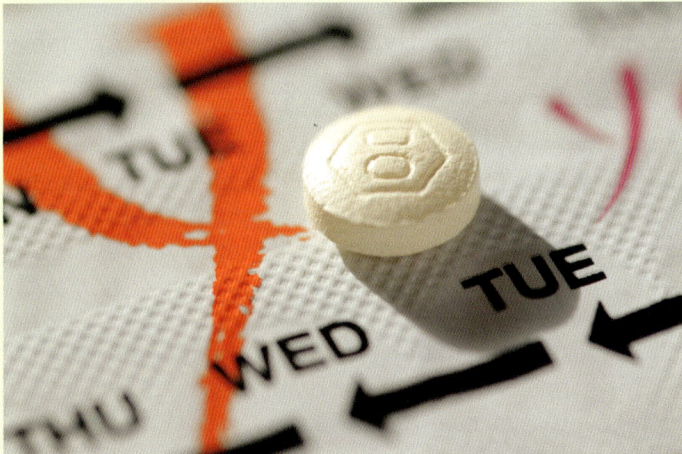

Figure 1.15 The contraceptive pill Laura was prescribed

❶ Why do you think Laura's doctor suggested the pill as a method of contraception? Discuss this in a pair and write a list of reasons for using the pill.

❷ What possible disadvantages are there in using the pill as a method of contraception? Use the following website to help you answer this question: www.fpa.org.uk. Click on the following links: 'Contraception and sexual health guide', 'Contraception', 'The combined pill', 'Are there any risks?'

❸ The pill contains the hormone oestrogen. Use Table 1.2 to explain how oestrogen can prevent an egg from maturing.

After three years Laura and David decided to start a family so Laura stopped taking the pill. They had sex regularly, but after a year Laura was still not pregnant. Her doctor referred them to a fertility clinic. After some tests the doctor found that the infertility was due to Laura's eggs not maturing properly in the ovaries. This was caused by a low concentration of one of her hormones. She was prescribed a fertility drug that contained a synthetic form of the hormone. The hormone caused her eggs to mature properly in the ovaries. After six months Laura became pregnant.

❹ Look at Table 1.2. What hormone do you think the fertility drug contained? Explain your answer.

❺ a) Some people do not agree with the use of synthetic hormones to treat infertility. Their reasons may be based on:
   A evidence, B hearsay, C prejudice, D personal opinion.
   Look carefully at the reasons given by the following four people for not agreeing with the use of synthetic hormones to treat infertility. In each case, decide which of A, B, C or D their reason is based on.
   i) Kelly says 'It's wrong to interfere with nature.'
   ii) Ali, her husband says 'The use of synthetic hormones could create problems in Kelly's hormonal system.'
   iii) Becky says 'It's against my religious beliefs.'
   iv) Lisa, her friend, says 'The synthetic hormones are drugs and all drugs are bad for you.'
  b) What is your opinion on the use of synthetic hormones to treat infertility?

In some cases of infertility taking this hormone does not help a couple have children. Another treatment is called IVF (*in vitro* fertilisation). This involves the woman taking another hormone to stimulate egg production. Eggs are then removed from her ovaries with a needle and fertilised with the man's sperm in a laboratory. A few of the fertilised eggs are then inserted into the woman's womb in the hope that one or more of them will implant and develop into a baby.

❻ What causes of infertility do you think IVF can be used to treat?

❼ Suggest two potential problems related to IVF.

How do our bodies maintain the right temperature?

Internal body temperature (°C)	Symptoms
28	Muscle failure
30	Loss of body temperature control
33	Loss of consciousness
37	Normal
42	Central nervous system breakdown
44	Death

Table 1.3 Symptoms that occur when the body's internal temperature changes

Look at Table 1.3. It shows that changes in your body temperature can have very serious effects on how it functions. Most of these changes take place because enzymes controlling the processes in your body work best at normal body temperature. This is 37 °C. When the temperature changes, enzyme molecules begin to change shape. If this happens they cannot catalyse body processes effectively.

Fortunately your body has a number of ways of controlling its temperature. It can cool itself down if it gets too hot and warm up if it gets too cold.

How does the body cool down?

Your body will warm up on a hot day or when you take exercise. Imagine a marathon runner on a hot day and you can probably picture how the body responds to its increasing temperature.

Sweating is often the first sign that your body is trying to reduce its internal temperature. Sweat is produced by sweat glands in the skin and secreted onto its surface. As water in the sweat evaporates off the skin, it takes heat away from the body. This reduces the internal body temperature.

When your skin becomes red on a hot day, it has more blood flowing near the surface. This happens because your brain sends a nervous impulse to the blood capillaries near the surface of your skin which causes them to dilate (widen). Some of the heat that the blood carries to the surface is then lost to the air. This response is called vasodilatation.

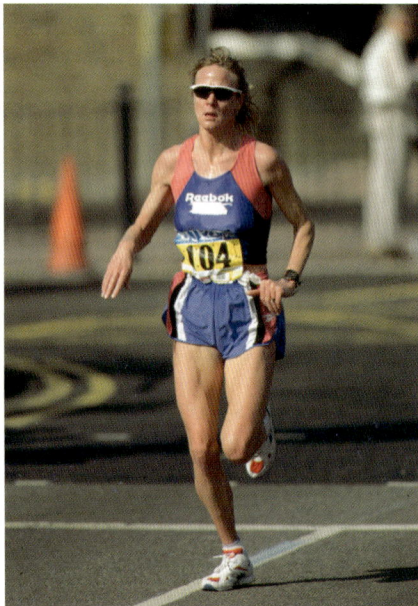

Figure 1.16 A marathon runner on a hot day

⑲ Look at the photo in Figure 1.16. How is the marathon runner's body responding to the increase in her body temperature?

⑳ How do you think these changes reduce her body temperature?

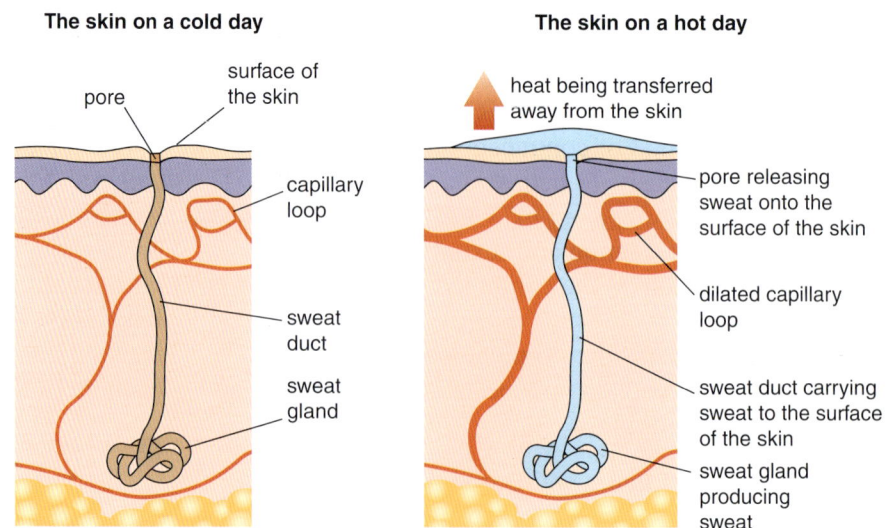

Figure 1.17 Changes that occur on the surface of the skin on a hot day

㉑ Explain how stopping sweating, vasoconstriction and shivering help to raise body temperature.

㉒ Hypothermia is a condition that occurs if body temperature drops below 35 °C. Many more old people than young people die from hypothermia. Why do you think this is? Explain your answer.

How does the body warm up?

When the temperature around you drops, your body temperature will start to fall. Your body will respond to this by retaining the heat that is being released during respiration. It does this by stopping sweating and by reducing the blood flow through the capillaries in the surface of the skin. This happens when the capillary loops in the skin (Figure 1.17) are closed off by a process called vasoconstriction. Your body will also produce more heat by shivering. Shivering causes your muscles to contract and relax very quickly. This causes respiration in your muscle cells and heat is produced.

1.4 How do drugs affect the body?

Drugs are substances that have an effect on processes in the body. Medicines are drugs. When used properly, they are beneficial to the body. Some people use substances like nicotine, alcohol, cannabis and heroin as recreational drugs. Some of these are more harmful than others. It is very important to know exactly how a drug affects our bodies before we can assess its possible risks or benefits.

Medical drugs

Medical drugs or medicines are used to treat disease, injury and pain. Medicines are beneficial when used properly but can still be harmful if misused. Paracetamol is an effective painkiller if you take the correct dose but taking more than this can cause liver damage. In extreme cases, it can kill. This is why it is essential that you always follow the instructions given with medicines.

There is evidence of medicines being used as far back as prehistoric times (8000BC). Cave paintings show tribal healers feeding people plants to cure illness. The Egyptians had doctors in 200BC who used herbal medicines to treat diseases. Some of the medicines used today contain chemicals that were present in these ancient treatments and many more are derived from natural substances found in plants today. Over 7000 chemicals used to make medicines are found in plants and many of these plants are found in the rainforests. The indigenous people of the rainforest use the plants growing there to treat many illnesses. Quinine is extracted from the cinchona tree, found in South America, and is used to treat malaria. A plant called rosy periwinkle, found in Madagascar, provides two of the most important chemicals used to treat cancer. New drugs are still being developed from natural substances found in plants.

Figure 1.18 A range of medical drugs available today

Figure 1.19 The cinchona tree a) and rosy periwinkle b) are both used to produce essential medicines

23 Why is it important to carefully follow the instructions when you take medicines?

24 Why do you think it is important to medical science that the rainforests are preserved?

25 What are the advantages of testing a new drug on live cells instead of on animals?

26 Why do you think new drugs are tested on healthy people instead of people with the disease in stage 2 of the testing process?

27 Why do scientists bother testing a new drug against an existing medicine in stage 4 of the process?

28 Describe the side effects of the drug thalidomide. Why do you think the side effects of thalidomide were not detected when the drug was tested?

When scientists develop a new drug to treat a disease it must be thoroughly tested. The testing is carried out in a number of stages.

Stage 1: Drugs are tested in the laboratory. They are tested on live cells and sometimes on animals.

Stage 2: Drugs that show promise in the laboratory are tested on a small number of healthy humans.

Stage 3: The drug is tested on about 100 humans who suffer from the disease that the drug is intended to treat.

Stage 4: The drug is tested on several hundreds of sufferers alongside an existing drug for the disease.

Only drugs that pass all safety checks and clinical trials can be launched as a new product. In the vast majority of cases medical drugs, when used as instructed, have no, or only limited, side effects. Very occasionally a drug that has been tested causes unexpected side effects. A drug called thalidomide was used in the 1960s as a sleeping pill. It was also effective at relieving morning sickness in pregnant women. Unfortunately, many babies born to mothers who took the drug had very underdeveloped arms and legs. The drug was then banned but has since been used successfully to treat leprosy.

The testing and development of modern medicines and drugs illustrates the way in which research scientists work. The process often begins with an idea or an explanation using existing theories as to why a particular drug or medicine might be used to treat a specific disease. This idea or explanation is called a **hypothesis**. The hypothesis can then be used to make predictions about the effects of the drug which can be thoroughly tested.

The tests involve laboratory experiments before any trials on humans.

If the experiments and observations support the initial hypothesis about the drug, then **clinical trials** can begin. These trials will use carefully selected groups to check the effects of the drug on people of different sex, age and weight. They will also involve control groups to ensure that the effects of *only the drug* (the independent variable) are observed. The control group will be given a placebo (tablets that look like the drug but have no effect). All those involved in the clinical trials will not be told whether they are having the experimental drug or the placebo.

If, after all these tests and trials, the results support the initial predictions about the drug, then large-scale production can begin.

Figure 1.20 A person suffering from the side effects of thalidomide

Activity – What does research tell us about the use of cannabis?

Recently there has been a lot of research into the use of cannabis. One area of research has investigated the possible link between taking cannabis and then moving on to hard drugs, such as heroin. Another area of study has tried to establish a link between smoking cannabis and health problems, especially psychological problems. There have been some conflicting findings.

Read the following quotes from different researchers.

i) 'Cannabis use does not lead to experimentation with harder drugs.'

ii) 'Cannabis does not act as a "gateway" drug to the use of harder drugs.'

iii) 'Many hard drug users have followed a similar path from cigarettes and alcohol, to cannabis, to heroin and cocaine.'

iv) 'Early marijuana smokers were found to be up to five times more likely to move to harder drugs than were their twins who did not smoke marijuana.'

v) 'Some studies have suggested long-term cannabis use can increase your risk of developing schizophrenia.'

vi) 'Smoking cannabis virtually doubles the risk of developing mental illnesses.'

vii) 'There was no proven causal link between taking cannabis and mental illness.'

❶ Find two pairs of totally opposite conclusions in the above quotes.

❷ Summarise the views expressed in the quotes in four sentences.

❸ Different researchers have clearly come to very different conclusions about the links between cannabis smoking, the use of hard drugs and mental illness. In a pair discuss why you think this is. Write a list of the reasons why researchers can come to such conflicting conclusions.

❹ Newspapers often use quotes from researchers in their headlines.
   a) What problems might this lead to given the range of opinions expressed by these quotes?
   b) What other information about research into cannabis smoking should be included in a newspaper article to give the public a clearer understanding of the issues?

❺ Plan a research project to investigate a possible link between cannabis use and mental illness. You will need to consider the following points.
   - Who will be involved?
   - How many people will be involved?
   - How long should the study last?
   - How will you include a control in the study?
   - How will you collect information from the people involved?
   - How will you analyse the results?
   - How will you share your findings with other people?

How does smoking tobacco affect the body?

Tobacco smoke contains more than 4000 chemicals, many of which are poisonous and could kill you. Three of the dangerous chemicals in tobacco smoke are nicotine, tar and carbon monoxide. Nicotine is an **addictive** substance which causes people to become **dependent** on tobacco smoke. Tar is actually a mixture of many chemicals which cause lung cancer, bronchitis and emphysema. Carbon monoxide reduces the blood's ability to carry oxygen around the body. It can deprive an unborn baby of oxygen whilst in the mother's womb and lead to a lower birth weight. In spite of all these health problems from the smoking of tobacco, cigarette manufacturers were not required to put government health warnings on cigarette packets until 1971.

Figure 1.21 Government health warnings on tobacco products

Have a look at the key dates below, relating to smoking and health.

1951: Dr Richard Doll and Prof Austin Bradford Hill conduct the first large-scale study of a link between smoking and lung cancer.

1954: Dr Doll and his team publish a paper confirming the link.

1957: The British Medical Research Council announces 'a direct causal connection' between smoking and lung cancer.

1962: A report from the Royal College of Physicians concludes that smoking is a cause of lung cancer and bronchitis, and probably contributes to coronary heart disease. The report recommends tougher laws on cigarette sales, cigarette advertising, and smoking in public places.

1965: The British government bans cigarette advertising on television.

1971: Government health warnings are required on all cigarette packets sold in the UK, following an agreement between the government and the tobacco industry.

1976: Prof Richard Doll and Richard Peto publish the results of a 20-year study of smokers. They conclude that one in three smokers dies from the habit.

1983: A report from the Royal College of Physicians features passive smoking for the first time. It also asserts that more than 100 000 people die every year in the UK from smoking-related illness.

1988: An independent Scientific Committee on Smoking and Health concludes that non-smokers have a 10–30% higher risk of developing lung cancer if exposed to other people's smoke.

1989: A UK court rules that injury caused by passive smoking can be an industrial accident.

29 a) How long did it take from the start of the first large-scale study into tobacco smoking until health warnings were put on cigarette packets?
   b) Why do you think this took so long?

30 Which of the dated developments do you think would have the greatest effect in persuading a smoker to give up smoking? Explain your answer.

31 Find more information about the diseases a) bronchitis and b) emphysema.
   i)   What causes the disease?
   ii) What are the symptoms?
   iii) How is the disease treated?

What other drugs may harm the body?

Drug	Classification	Effects on the body
Alcohol	• Illegal for anyone under the age of 18 to buy • Illegal for someone over 18 to supply to an under 18	• Slows down reactions • Feel relaxed • Lose inhibitions • Can become loud and sometimes aggressive • Unconsciousness • Increased blood pressure • Possible liver and brain damage
Solvents (glue, lighter fluid, paint thinners, correcting fluid)	• Illegal for a retailer to sell a solvent to anyone under the age of 18 if they believe it will be used for inhaling to cause intoxication	• Lose inhibitions • Blurred vision • Dizziness • Blackouts • Possible lung, liver and brain damage
Cocaine	• Class A or 'hard' drug. Illegal to possess or sell	• Feeling of well-being and confidence • Increased heart rate and blood pressure • Highly addictive • Depression • Mental health problems
Heroin	• Class A or 'hard' drug. Illegal to possess or sell	• Feeling of well-being • Drowsiness • Blurred vision • Vomiting • Highly addictive • Can cause a coma or death when taken in large amounts

Table 1.4 The classification and effects of some drugs

32 Research three different approaches that people use to stop smoking. For each method a) explain how it helps a smoker to give up smoking, b) suggest an advantage and a disadvantage for the approach.

33 List the short term and long-term effects of a) alcohol; b) cocaine.

34 Treatment for drug addicts is freely available on the National Health Service.
a) State three points in support of this policy.
b) State one point which criticises this policy.

35 Suggest three things that could be done to reduce the number of drug addicts in the UK.

Summary

✓ The **nervous system** enables humans to react to stimuli from their surroundings and co-ordinate their behaviour.

✓ The nervous system includes: **receptors, neurones,** the spinal cord, the brain and **effectors** (muscles and glands).

✓ Receptors detect stimuli which include light, sound, changes in position, chemicals, touch, pressure, pain and temperature.

✓ Information from receptors passes along neurones, in the form of nervous impulses, to the brain. The brain co-ordinates the response.

✓ **Reflex actions** are automatic and take place very quickly. They involve sensory, relay and motor neurones.

✓ Chemicals called **hormones** are involved in controlling conditions inside the body. These conditions include the water content of the body, the ion content of the body, body temperature and blood sugar levels.

✓ Hormones are secreted by glands and are transported by the bloodstream to their **target organ**.

✓ Several hormones are involved in controlling the menstrual cycle. These hormones include FSH, oestrogen and LH.

✓ Hormones can be used to control **fertility**. This includes using oral contraception to stop a woman's eggs maturing and prescribing FSH as a fertility drug to stimulate eggs to mature.

✓ **Drugs** are chemicals that have an effect on processes in the body.

✓ **Medicines** are drugs that, when taken properly, are beneficial to people. Other drugs can harm the body.

✓ Many drugs are derived from natural substances.

✓ When new drugs are developed they must be thoroughly tested. New drugs are tested in the laboratory and then on human volunteers.

✓ Occasionally drugs that have been tested cause unexpected side effects. Thalidomide is a drug that caused deformed limbs in the children of mothers who took the drug.

✓ Some drugs that are used recreationally are very addictive. Heroin, cocaine and nicotine are examples of addictive drugs.

✓ Claims have been made about the effects of cannabis on health and a link between taking cannabis and addiction to hard drugs. However, research has led to conflicting conclusions about these claims.

✓ The link between smoking tobacco and lung cancer took some time to be accepted.

✓ Tobacco smoke contains a number of chemicals that damage health.

✓ Alcohol affects the nervous system by slowing down reactions. It does help people to relax but too much can lead to lack of self-control and even unconsciousness. Alcohol can also cause liver and brain damage.

1 A student accidentally touches a hot saucepan. His hand automatically moves away from the pan.

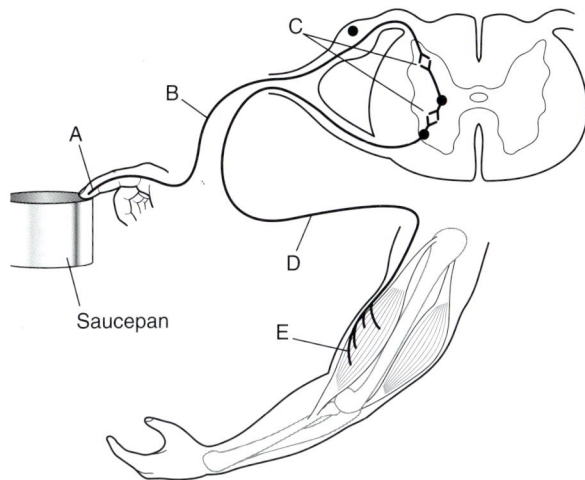

Figure 1.22 The different elements involved in a reflex action.

a) In this reflex action:
   i) where is the receptor?
   ii) where is the effector? *(2 marks)*
b) Explain how an impulse crosses the synapse labelled C. *(1 mark)*
c) Explain why this type of reflex action is important to our bodies. *(1 mark)*

2 A hormone is involved in controlling the water content of the body.
a) State **two** ways in which water leaves the body. *(2 marks)*
b) What can cause a rapid fall in the water content of the body? *(2 marks)*
c) Explain how the body responds to a fall in water content. *(3 marks)*

3 Read the information about contraceptive implants.

A contraceptive implant works a bit like a contraceptive pill. They contain a hormone present in contraceptive pills, which prevents pregnancy. The hormone implant is a small, thin flexible rod, 4 cm long and made of plastic. It is inserted just under the skin of a woman's arm. This procedure must always be undertaken by a doctor who is familiar with the technique.

The implant gradually releases a small amount of the hormone, which prevents pregnancy for up to three years.

a) How can a hormone prevent pregnancy? *(2 marks)*
b) Give **one** drawback of using contraceptive implants rather than contraceptive pills. *(1 mark)*
c) Contraceptive implants are being used increasingly in birth control programmes in developing countries, instead of contraceptive pills.
   Can you think of a reason for this? *(2 marks)*

4 A person's blood sugar level rises after a meal. Following this the pancreas secretes a hormone called insulin. This causes the liver to store sugar until it is needed.
a) i) Name the gland involved in this response. *(1 mark)*
   ii) Name the target organ involved in this response. *(1 mark)*
b) Explain how insulin is transported from the pancreas to the liver. *(2 marks)*
c) Why does the liver need to release sugar when a person is taking vigorous exercise? *(1 mark)*

5 Medicines must be thoroughly tested before they can be prescribed to the public.
a) Outline the procedure for testing a new medicine. *(2 marks)*
b) A drug called thalidomide caused severe side effects despite passing safety tests and clinical trials.
   i) Describe the side effects caused by thalidomide. *(2 marks)*
   ii) Explain why thorough testing did not identify the possibility of these side effects. *(3 marks)*

6 Alcohol and tobacco can damage the body.
a) Give one example of how the body is damaged by tobacco smoke. *(1 mark)*
b) Why do you think that people find it difficult to give up smoking? *(2 marks)*
c) Why should motorists not drive after they have been drinking alcohol? *(2 marks)*

Chapter 2
What can we do to keep healthy?

At the end of this chapter you should:

✓ understand the importance of different food groups for your health;

✓ be able to discuss the different types of fat and their effect on cholesterol level;

✓ understand the link between energy and specific nutrient requirements with age;

✓ know how to calculate the percentage of saturated fat in food;

✓ appreciate the link between exercise, food energy and health;

✓ know how to calculate the amount of energy used with exercise;

✓ be able to evaluate whether or not a diet is healthy;

✓ be able to explain the body's defences against pathogens;

✓ appreciate the importance of immunisation programmes;

✓ be able to assess the value of and problems associated with antibiotics.

Figure 2.1 Food can be colourful and interesting

2.1 Eating for health

> **1** a) List eight different coloured fruits or vegetables. Try to include all the colours of the rainbow.
> b) Try to name all the fruits and vegetables shown in Figure 2.1.

Do you eat the same foods every day? Do you think about your choice of food? Most people eat a varied diet that includes everything needed to keep their body healthy. The important word in the last sentence is 'varied' because no single food can provide all the essential nutrients your body needs to function efficiently. An easy way to ensure that your diet contains all the essential vitamins and minerals is to make sure you eat as many different naturally coloured foods as possible. Red tomatoes, green broccoli and purple plums are three examples.

For a healthy diet you should aim to eat some food from each of the groups in Table 2.1 every day.

Food group	Examples	Health points
Energy foods	• Whole grains such as rice, millet • Yams, potatoes • Cereals, bread and pasta	• Whole grains provide extra fibre, which means starchy foods are converted and released as sugars over a longer time period, they are low GI foods. This helps to prevent snacking
Food for building bones and teeth (two or three servings daily)	• Dairy products such as milk, cheese, yoghurt, fromage frais • Soya or rice milk with added calcium	• Skeletons increase in strength most rapidly in the late teens • Skimmed and semi-skimmed milk are as calcium-rich as other milk but have less fats
Muscle-building foods	• Red meat, poultry, sausages, bacon, eggs • Nuts, tofu, beans, lentils, quorn (mycoprotein) • Fish such as salmon, sardines, cod	• Different protein foods are important for vegetarians to ensure they get all the essential **amino acids**. Red meat eaters should avoid too much excess fat • Oily fish such as sardines and salmon provide the **essential fatty acids** (EFAs) which are good for the brain • Meat and fish contain vitamins A, D and E

Foods for clear skin and a healthy digestive system. (A minimum of two portions of fruit and three of vegetables daily)	• Leafy vegetables such as cabbage and broccoli
	• Peas, beans, courgettes, squash, okra, onions
	• Salad such as lettuce, cucumber, peppers, tomatoes
	• Fruit such as oranges, nectarines, strawberries, mangoes, blueberries
	• Dried fruit such as apricots, prunes
		• The brightly coloured pigments in fruit and vegetables contain minerals and vitamins
		• Fruits are rich in vitamin C
		• The fibre in fruit and vegetables:
		  – speeds the passage of food through the intestine helping prevent colon cancer;
		  – slows the release of sugars, which helps to reduce the risk of diabetes and prevent snacking

Table 2.1 The health points of different food groups

Figure 2.6 A little of what you fancy does you good. Too much makes you fat.

Amino acids are molecules that build up into proteins.

Fatty acids are part of fat molecules.

❷ What health problems might a child start to experience if they refused to eat fruit and vegetables for a few weeks?

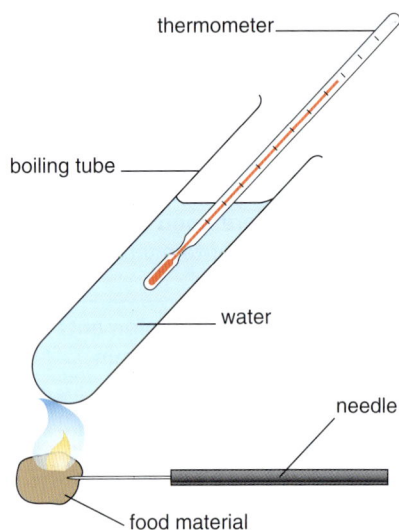

thermometer

boiling tube

water

needle

food material

Figure 2.7 Apparatus used to measure the energy value of different foods

If you don't eat all these essential nutrients you will become malnourished whatever your weight. The eating suggestions listed in Table 2.1 contain all the vitamins and minerals you need. But your meals must give you the correct amount of energy or the correct number of joules.

How do we compare the energy content of food?

Energy values are quoted on the 'nutritional information' panels on food packets. In some cases the old units, calories, are still used.

The following experiment will help you to understand how manufacturers obtain the energy value for a food product.

If you were to carry out an experiment to compare the energy values of foods, you might take pieces of food, set fire to them, hold a boiling tube containing water above the flame, and see which heats the water the most.

How many factors have not been controlled in this suggestion and the apparatus shown in Figure 2.7? Think about this before you look at the hints below.

Did you consider the following points?

❶ There should be only one independent variable and for this experiment we are trying to compare the energy values of different foods. Everything else must stay the same (**control variables**) for each different food.
How could you make the mass of food a fair test?

❷ Have you tried to burn foods? It can take quite a lot of heat to start the reaction in air. If you heated the foods over a Bunsen burner for different times this would be an additional variable. Can you think of another method of getting the food to burn?

❸ Do substances burn better in oxygen than air?

❹ What about the volume of water? Should the water be stirred to distribute the heat?

❺ Will all the heat go into the water in a boiling tube? If not, where will it go? Could you reduce heat losses? Why could this be described as a **systematic error**?

❻ The boiling tube is made of glass. Is glass the best material to use or could you suggest a better conductor of heat?

❼ How would you measure the temperature rise of the water?

After you have evaluated this experiment, look at the diagram of the food calorimeter in Figure 2.8 and see how many of your criticisms have been overcome.

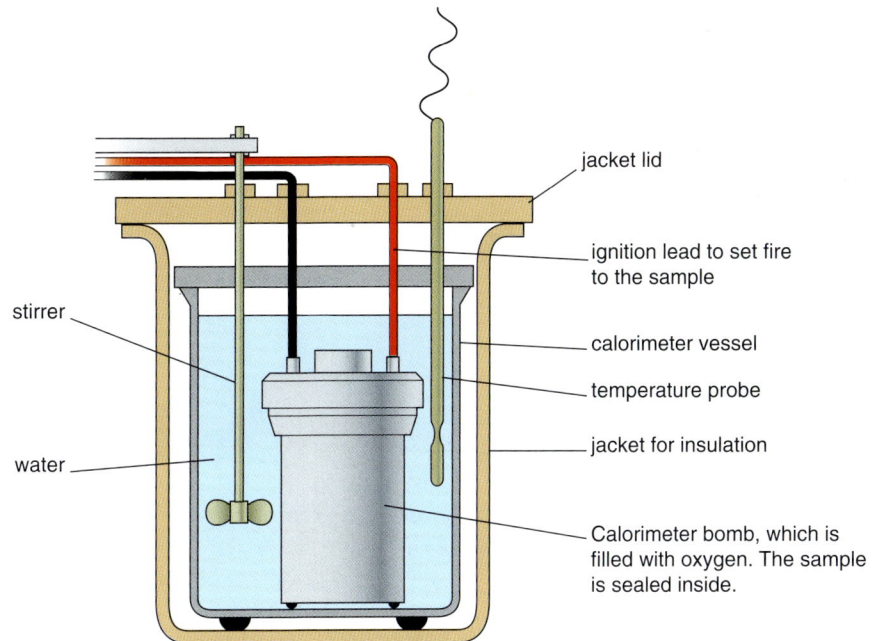

jacket lid

ignition lead to set fire to the sample

stirrer

calorimeter vessel

temperature probe

jacket for insulation

water

Calorimeter bomb, which is filled with oxygen. The sample is sealed inside.

Figure 2.8 A food calorimeter. Experiments with food calorimeters show that 'one gram of fat contains more than twice as much energy as one gram of carbohydrate'

Sorting out the fats
The terms saturated, monounsaturated and polyunsaturated are used to describe different fats.

Saturated fats are hard at room temperature. They include fat around meat or left in the pan after cooking.

Unsaturated fats are more likely to be liquid at room temperature. The word 'unsaturated' is used to describe the double bonds between some carbon atoms in their structure.

Should we avoid eating all fats?

A high-fat diet can lead to obesity. Read the following facts and then decide if we should avoid eating all fats. Fats have important functions in the body:
* they protect vital organs from damage;
* they form an energy reserve;
* essential fatty acids (EFAs) are used in our bodies to make hormones, including the sex hormones;
* they contain cholesterol which is important for making strong cell membranes;
* fats contain fat-soluble vitamins (A, D, E).

Look on the side of a food packet and you will find the 'nutritional information' which has a list of the fat types present.

Both mono- and polyunsaturated fats help to reduce cholesterol, but the polyunsaturated fats contain EFAs. Figures 2.9–2.11 will help you to picture these different fats. Saturated and unsaturated fats are studied more fully in Sections 7.4 and 7.5.

Figure 2.9 Dairy products and red meats contain saturated fats. Saturated fats can be used for energy but excess is stored in the body as fat deposits. Too much saturated fat can raise your blood cholesterol level.

Figure 2.10 Olive oil, peanuts and avocados contain monounsaturated fats. These can be used for energy and help to lower blood cholesterol levels.

Figure 2.11 Oily fish, sunflower seeds and vegetable oils contain polyunsaturated fats. They also contain essential fatty acids called omega-6 and omega-3 which help to lower the blood cholesterol level. Even though the EFAs have essential roles in the body excess is still stored in the fatty tissues.

❸ Nutritionists say that saturated fats should be limited to about 10% of the daily recommended energy intake for an active teenager. Why do they give this advice?

❹ Nutritionists also suggest that we should eat oily fish, such as salmon or sardines, at least twice a week. Why do they give this advice?

❺ Although there are more and more people who are obese, the amount of energy in the food that people eat has decreased slightly. Suggest a possible reason for these contradictory statements.

Fats are sometimes combined in molecules with proteins. These molecules are called lipoproteins. Lipoproteins are important in our bodies because they transport cholesterol in the blood. High-density lipoproteins (HDLs) carry cholesterol out of the blood vessels and this is sometimes called 'good' cholesterol. Low-density lipoproteins (LDLs) are responsible for depositing cholesterol in blood vessels and so this is sometimes called 'bad' cholesterol.

Cholesterol is important in the body for making strong cell membranes and vital hormones. The cholesterol level in the blood depends on:
- the amount produced by the liver (an inherited factor);
- the amount eaten in food.

If the cholesterol level in the blood is too high, fat can be deposited in the lining of blood vessels. This raises blood pressure and increases the risk of heart disease. Women have a high level of HDLs until the menopause. The HDLs transport cholesterol out of blood vessels to the liver to be broken down. HDLs protect younger women from fatty deposits. Men have lower levels of HDLs and higher levels of LDLs so they are more at risk of blocked blood vessels, especially if they eat too much saturated fat and also smoke.

6 What sort of foods contain cholesterol?

7 Some diets are advertised as 'cholesterol free'. Is a cholesterol free diet a good idea? Give the reasons for your answer.

8 Earlier in this chapter we raised the question 'Should we avoid eating all fats?' Give at least five reasons why your answer should be 'No'.

9 Look carefully at the following statements. Which do you think is the best explanation of childhood obesity? Explain your choice.
a) 'Children get fat because they watch too much television or play too many computer games.'
b) 'Children get fat because they watch television rather than take part in active outdoor games.'
c) 'Children get fat because they snack all the time while watching television.'

The nutritionist's viewpoint
Nutritionists have come to the conclusion that 'obesity is a complex problem caused by the interaction of many factors' and there is no single 'quick-fix' solution.

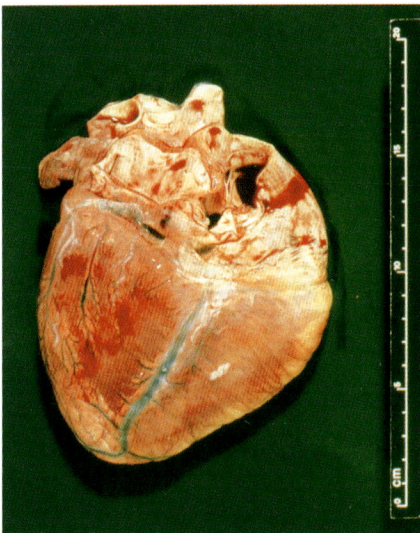

Figure 2.12 Save your heart! Don't add salt to food at the table or in cooking. Be more adventurous and add herbs for flavouring.

Using 'nutritional information' on food packets

Nutritionists recommend that the proportion of saturated fats in our diet should be about 10%. We can use the nutritional information on food packaging to check this.

A small 50 g bar of chocolate contains 30 g fat per 100 g chocolate.
So, the mass of fat in the chocolate bar = 15 g.
(If saturated fats are not listed separately a reasonable estimate is half of total fats.)
Therefore, the mass of saturated fats in the chocolate bar = 7.5 g.

1 g fat provides 37 kJ of energy.
Therefore, the saturated fats in the chocolate bar provide
$7.5 \times 37 = 277.5$ kJ of energy.

The average female teenager requires 8100 kJ of energy per day and 10% of the energy should be saturated fats – this is 810 kJ.

So, the small chocolate bar provides $277.5 \div 810 \times 100\% = 34\%$ of the daily recommended total of saturated fats.

Salt in our diet

Salt is added to food as a preservative and to 'bring out the flavour'. It is very easy to exceed the recommended salt intake. Children in particular are at risk of consuming too much salt if they eat ready meals and crisps. There is very little salt in raw or fresh foods. The salt intake for a 7–14 year old should not exceed 5 g per day and for over 14s and adults it should not exceed 6 g per day.

Doctors are concerned about the amount of salt that people have in their diet. High levels of salt are linked to high blood pressure, which is a risk factor for heart disease and strokes. Heart disease is a major cause of death in the UK.

Activity – What is the proportion of fat in your diet?

a) Keep a food diary for one day. Write down exactly what you have eaten after each meal or snack because it is easy to forget.

b) Have a look at the 'nutritional information' on the packaging of the foods you eat. Alternatively you could check them on the internet. Find the amount or type of fat contained in your food.
Construct a table using the following headings. You will need one row for each different food eaten.

Mass of food eaten (g)	Mass of fat (g)	Mass of saturated fat (g)

Table 2.2 My food diary

Use the information in the section above to complete the following activities.

c) Complete your food diary and calculate your total mass of fats and saturated fats.

d) Does your saturated fat intake exceed 10% of your recommended energy intake? (Remember that the recommended energy intake for a female teenager is 8100 kJ per day and for a male teenager is 10 000 kJ per day.)

e) Which food gives you most saturated fat?

f) What problems might you have if your total fat intake is too low?

g) What foods should you eat only in limited amounts if you want to 'eat for health'?

h) Suggest five ways to reduce your fat intake.

How do our energy and nutrient requirements change with age?

Figure 2.13 Pre-school children are small but growing and they are very active. They need a varied diet which provides:
- enough energy for their activity;
- a good supply of protein for growth;
- calcium for bone formation;
- iron for haemoglobin in the blood;
- the fat-soluble vitamins A and D for general good heath.

The minerals and fibre should come from fruit and vegetables. They should eat small regular meals because of their small stomach size.

Figure 2.14 Pre-teens should:
- eat food from all categories listed in Table 2.1;
- eat more starchy foods than pre-school children, such as wholegrain cereals and potatoes for energy;
- avoid too much fatty food such as cakes, biscuits, chocolate and ice-cream.

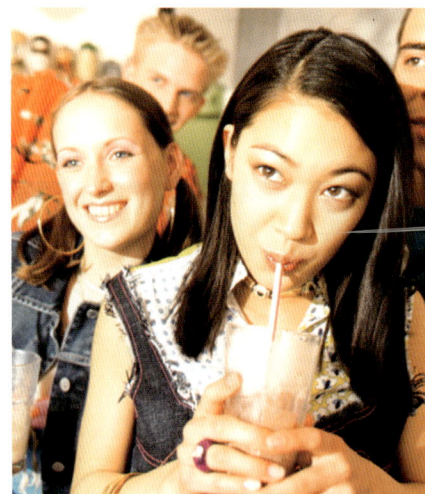

Figure 2.15 Teenagers should be aware of what they are eating. The teenage growth spurt should be supported by:
- an increase in protein foods;
- an increase in iron for male muscle building and female menstruation (periods);
- sufficient calcium-rich foods for bone development;
- 'five-a-day fruit and vegetables' to provide minerals and vitamins for general good health and clear skin;
- drinking milk (semi-skimmed rather than full-fat milk to reduce fat intake).

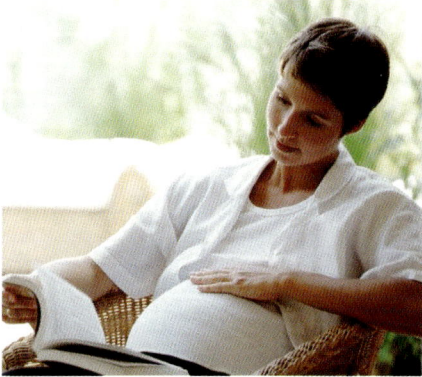

Figure 2.16 Pregnant and breast-feeding mothers are eating for themselves and a baby (not for two adults). They therefore need:
- more calcium for bones and teeth;
- a little more protein;
- fruit and vegetable to give them plenty of vitamins and minerals.

Figure 2.17 Adult diets should be related to exercise. For example, a window cleaner will need to eat more than an office worker. Adults need to eat a balanced diet including all food groups but avoiding excess fat, sugar and alcohol.

Figure 2.18 Older people over 65 tend to reduce the variety of their food and suffer from vitamin and mineral shortages. They should:
- continue to eat a range of fruit and vegetables;
- eat some protein every day;
- change their energy intake to balance the exercise they take.

2.2 Exercise for health

Activity	Energy used in joules (J)
Walking slowly	314
Walking quickly	628
Pushing electric/petrol lawnmower	700
Cycling at normal speed	750
Aerobics	810
Swimming (steady crawl)	820
Tennis doubles	630
Tennis singles	1000
Running 1 km in 6.2 minutes	1250
Running 1 km in 4.7 minutes	1700

Table 2.3 The energy requirements of different activities
The table shows the energy a person weighing 60 kg (approx 9.5 stone) would use in 30 minutes of different activities.

More energy is used if:
- your weight is greater;
- you are fit with well-developed muscles.

10 a) Which foods should a pregnant woman increase in her diet?
   b) What is the reason for each increase?

11 a) If you eat several bars of chocolate and bags of chips what might you be taking in excess?
   b) What health problems might result if this habit continued for a long time?

12 Heart disease is a problem in the UK. What could you do to reduce your risk of heart disease?

13 Haemoglobin in red blood cells transports oxygen. Which mineral is needed for the production of haemoglobin?

14 Look at the figures in Table 2.3. How does energy use change with speed?

15 a) Copy the table below and calculate the total energy used by Helen in a typical day. Assume her body mass is 60 kg.

Activity	Time (min)	(Show your working here)	Energy used (kJ)
Helen cycles to school and home again taking 7.5 minutes each way	15	Energy used in 30 minutes of cycling is 750 kJ Energy used in cycling for 15 minutes = 750 ÷ 2	375
On arrival she does one hour of swimming training			
At lunchtime she walks around slowly while chatting	15		
In the evening she plays tennis (doubles) for an hour			
		Total	

Table 2.4 Helen's activity record

b) Copy the table below and calculate the total energy used by Saroj in a typical day. Assume that his body mass is 60 kg.

Activity	Time (min)	(Show your working here)	Energy used (kJ)
The day starts with Saroj's paper round. This takes him 30 minutes walking quickly			
He is late so he runs the 2.0 km to school getting there in 12 minutes			
He rests at lunchtime		0	0
He walks home slowly, taking an hour to get there			
In the evening, he earns more pocket money by mowing an elderly person's lawn. He takes 45 minutes using an electric push mower			
		Total	

Table 2.5 Saroj's activity record

c) Now make a similar table to calculate your own energy use on a typical school day. How could you increase your exercise level?

Metabolic rate is the rate at which food is broken down chemically in our bodies. It can be measured in joules per minute.

Muscle tissue has a higher metabolic rate than fatty tissue so a muscular person uses up more energy than a skinny or fat person.

Environment or genetics?

Our diets, the exercise we take and the **metabolic rates** of our bodies are affected by genetic and environmental (lifestyle) factors. Humans evolved as active beings who were able to survive in an environment where food was sometimes in short supply. Our ancestors were hunter-gatherers and farmers. They had a good **energy in / energy used** balance. Today's car trip to the supermarket rarely finds any food shortages and does not use up much of our energy intake. In many parts of the world our lifestyle has changed significantly over the years.

- We use a car or bus to get around, rather than walk.
- Advertising companies urge us to eat high-energy foods such as crisps and canned drinks.
- Many people have desk-based jobs.
- We spend hours playing computer games, watching TV and DVDs – 'couch potatoes' and 'mouse potatoes' don't use much energy.

So compared with our ancestors we have a double disadvantage – **more energy (food) in and less energy used (exercise)** in our daily lives. The consequence of this is gain in weight.

How can we achieve the *energy in = energy used* balance to prevent weight gain?

In order to balance *energy in with energy used* and prevent weight gain we should:

- take regular exercise;
- be aware that cooking methods (such as frying) increase the energy in foods;
- know which foods contain hidden fats and sugars that provide surplus energy.

Figure 2.19 Balancing our food energy intake with the energy we use in our lives. Energy in = energy used

The benefits of exercise

Most gyms record a great increase in members around January 1st! Many people start with lots of enthusiasm but then stop attending as regularly as they did. Why do you think this is?

Ten good reasons for exercise are given on page 28 but it is important to remember that exercise needs to be both regular and enjoyable. You won't develop a 'six-pack' overnight, but firm well-toned muscles are well worth the effort.

Exercise is beneficial because it:

- increases the strength of your heart muscles and reduces blood pressure;
- improves lung ventilation (more air moved in and out of the lungs);
- improves oxygen uptake into the blood, which benefits your brain and body;
- increases muscle size and efficiency;
- strengthens ligaments and tendons and reduces the risk of injury;
- increases bone density and reduces the risk of osteoporosis (bone wastage). This disease affects both men and women but is more common in women after menopause. It can lead to fragile bones that break easily;
- decreases cholesterol levels in the blood and reduces the risk of blocked blood vessels;
- improves resistance to infections;
- raises metabolic rate and continues 'to burn up food' after you have finished exercising;
- leaves you with a 'feel good factor'.

If you check back, you can see that exercise and diet work together on improving the health of the whole body.

Activity – Food, health and exercise

1. Working as a group, check through a typical day and list:
   a) the high-energy foods you ate or avoided;
   b) the activity opportunities you took or avoided.
2. Is your life healthy or could you still make some positive changes for a healthier lifestyle?

Remember, any changes must be fun if you are going to keep them up.

3. Doctors say that little changes such as walking up stairs, taking a slightly longer route to walk home and using a bike instead of getting a lift are more beneficial than big ideas such as 'I'm going to jog five miles three times a week.' Can you suggest some reasons why?

2.3 What happens when the energy balance is upset?

A big problem
Obesity is a global health issue with 310 million people affected.

energy in

energy used

Figure 2.20 Energy in is greater than energy used

Energy in food eaten is greater than energy used. The calculations in question 15 show a direct link between exercise and energy used. Just look back at your conclusion. The key facts are:

i) the less exercise you take, the less food you need;

ii) if you take in more energy than you use, the excess is stored as fat.

In recent years, a lot of publicity has been given to obesity and the health problems it causes. In the past these were considered problems of 'old age' but now the same problems have been found in overweight teenagers and young adults.

Someone is regarded as overweight if they have a body mass index (BMI) greater than 25 and obese if they have a BMI of 30 or more.

Body mass index (BMI) is defined as $\dfrac{\text{body mass (kg)}}{\text{height}^2 \text{ (m}^2)}$

Body mass index (BMI) only gives a rough guide because it applies to over 18 year olds.

BMI does not take account of:
- the fact that muscle is denser than fat so athletes and body builders may have a BMI above 25 and still be healthy;
- variations in bone mass.

What are the problems of being overweight?

Overweight people are much more likely to suffer from arthritis, diabetes, high blood pressure and heart disease. These were once considered diseases of the 'over 50s'. Since the 1990s rates of these illnesses have been increasing. Some diseases, such as diabetes and rising blood pressure, are even seen in teenagers and young adults who are overweight. Children today consume the same food energy as children did 10 years ago, but they take far less exercise and so their energy input and energy use are not balanced.

The number of children with diabetes is increasing. Most of these diabetics take insufficient exercise (exercise uses up blood sugar) and 95% of them are overweight.

With such a high percentage, it is clear that there is a link between:
- diabetes and being overweight;
- diabetes and not doing enough exercise.

It is not surprising then that the treatment of diabetes in young people is to stick to a planned diet and do lots of exercise.

Diabetes in young people is taken very seriously because of increased blood pressure which results in heart disease and damage to nerves, the retina of the eye and the kidneys.

Figure 2.21 'I don't enjoy football, I can't run as fast as the others'

Being overweight also puts extra pressure on the joints, particularly those at the knee. This results in worn joints or arthritis, which can become a painful problem well before middle age.

What are the problems of being underweight?

If the *energy we take in with our food is less than the energy we need for our lifestyle,* this can also cause problems.

energy in　　　　　　　　　　　　　　*energy used*

Figure 2.22 Energy in food intake is less than energy used

You have probably heard of anorexia. This is a condition in which someone controls their own food intake so that it is well below that needed for healthy living. Anorexia affects about 1% of females. It is the third most common long-term illness in teenagers with a high mortality rate.

People also become underweight when there are food shortages. This often occurs when drought, famine or flooding hit developing countries. In these circumstances, people suffer from:
- deficiency diseases as a result of shortages of vitamins, minerals and essential amino and fatty acids (see Section 2.1);
- malnutrition as they are short of protein. Children are affected more than adults and have swollen bellies caused by water retention;
- a low resistance to infection as they are unable to form antibodies in response to pathogens (see Section 2.6). Epidemics such as measles spread rapidly through refugee camps;
- protein and iron shortages (in women). Their bodies respond by stopping periods, which helps them retain iron and protein and they can't become pregnant;
- mothers struggle to breast-feed their babies and so feed them on low-energy starchy foods with very little protein.

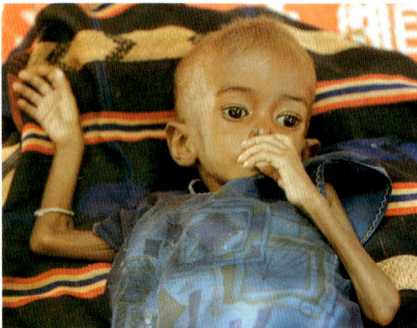

Figure 2.23 This toddler is suffering from malnutrition. The child has muscle wastage, no fat beneath the skin and is weak and lethargic.

The problems are made worse if the water supplies are contaminated with sewage. Water-borne diseases, such as cholera, are transmitted by drinking or cooking with contaminated water. Cholera causes diarrhoea and results in dehydration which is dangerous for babies.

2.4 Slimming

People who are slimming try to **reduce their energy intake so that it is less than the energy used**. Their aim is to lose weight and have an attractive well-toned figure. The best way to achieve this is to eat wisely and to exercise more. Although slimming is now an outdated term, it still represents a multi-million pound business. Slimming diets have now been replaced by healthy eating and fitness plans.

Any diet should be examined to ensure that it provides a wide variety of foods including all the nutrients required for good mental and physical health. All the food groups in Table 2.1 should be included, particularly fruit and vegetables. Any diet based on a single food, such as bananas or eggs, is not healthy in the long term. The aim must be to achieve a healthy eating and exercise plan so that weight remains constant with a body mass index (BMI) of between 20 and 24.

Activity – Evaluating slimming diets

In groups:
a) Collect three different slimming diets from magazines or books.
b) Evaluate these diets carefully using the following questions for guidance. You could produce a tick chart in response to each of the questions and then add up the ticks at the end.
c) What is the best way of presenting your findings to the rest of the class?
   - Will all the food groups listed in Table 2.1 be eaten each day?
- Is there the correct proportion of saturated fats?
- Will the essential fatty acids be eaten?
- Will the diet provide five servings of fruit or vegetables each day?
- Are the foods colourful and 'fun' to eat or are they boring and the same each day?
- Will the foods be satisfying and prevent feelings of hunger?
- Is the quantity of salt less than 6 g per day?
- Is exercise included?

In the last four sections we have seen that what we eat and how active we are can affect our health. In the next four sections we will look at the topic of disease.

2.5 What causes disease?

Diseases are caused by **pathogens**. Pathogens are microorganisms which include bacteria, viruses and fungi. They are called microorganisms because they are incredibly small and can only be seen with a microscope. Different pathogens invade different parts of the body. This helps doctors to identify which disease we have caught.

The smallest microorganisms are viruses.

> **Pathogens** are microorganisms (bacteria or viruses) which cause disease.

Viruses:
- are taken into cells of the body;
- use the cell contents to replicate and form thousands of identical copies;

Replicate means 'form an exact copy'. As viruses are not able to live independently they are not classed as 'living' so we do not use the word reproduce. They replicate using the DNA of the host cell. The damage to the host cell causes it to die.

- damage the cell as they burst out;
- infect nearby cells and repeat the process.

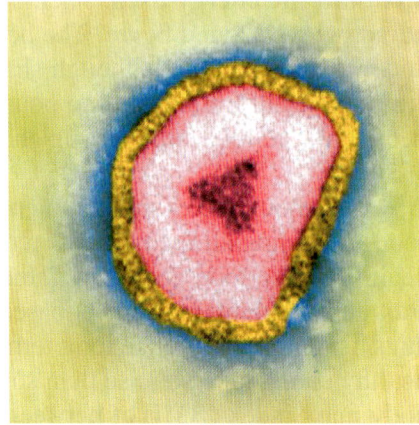

Figure 2.24 Photograph of virus taken using an electron microscope

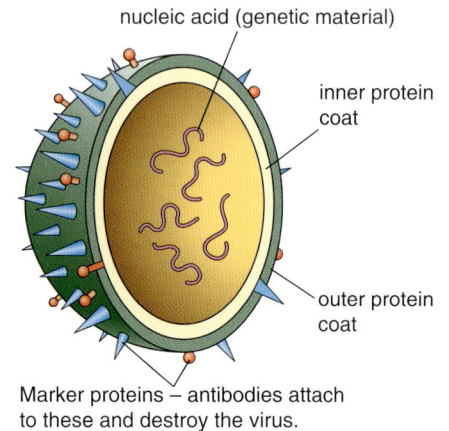

nucleic acid (genetic material)

inner protein coat

outer protein coat

Marker proteins – antibodies attach to these and destroy the virus.

Figure 2.25 The structure of a virus

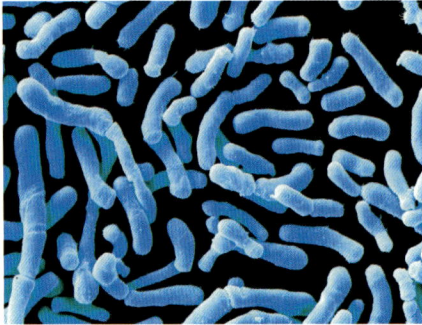

Figure 2.26 A photograph of bacteria taken with an electron microscope

Bacteria are much larger than viruses but can only be seen using a microscope. Pathogenic bacteria live in parts of the body such as the nose and throat but not inside cells. Some bacteria produce waste products which act as toxins and irritate cell membranes. These can cause sore throats and runny noses.

Hygiene

We learn to wash our hands before eating or cleaning a cut when we are very young. But, it hasn't always been so. In the 1800s many women died of blood poisoning following childbirth.

Ignaz Philipp Semmelweis, who was a doctor at a poor-mothers Maternity Hospital in Vienna, made the following observations:

- 29% of mothers died in the ward which doctors ran compared with only 3% of mothers in the ward run by midwives;
- few mothers died when giving birth at home;
- admissions after birth led to fewer deaths;
- only doctors carried out post-mortems (investigations to find the cause of death);
- doctors did not wash their hands between patients or after post-mortems;
- a doctor who cut his hand at a post-mortem died of the same blood poisoning as his patients.

Figure 2.27 Ignaz Philipp Semmelweis – an observant doctor

⑯ What do you think caused the difference in the death rate in the two wards?

⑰ Which are larger, bacteria or viruses?

Semmelweis ordered all doctors to wash their hands between patients. They objected at first, but then saw the death rate drop to 1% and realised he was right. In 1861 Semmelweis published his findings which were supported by the data he had collected. Many of the older doctors and scientists ridiculed his ideas and he died in 1865 before his simple lifesaving procedure became an accepted practice. (See MRSA in Section 2.7.)

2.6 The body's natural defence against pathogens

Antibodies are proteins produced by white blood cells that destroy pathogens which cause disease.

⓲ How does an antibody 'fit' onto a pathogen?

⓳ After the antibodies have caused the pathogens to stick together in a cluster, how are they destroyed?

⓴ If a person develops an infection why does it take a few days before they start to get better?

㉑ Why is a small child unlikely to catch chickenpox a second time?

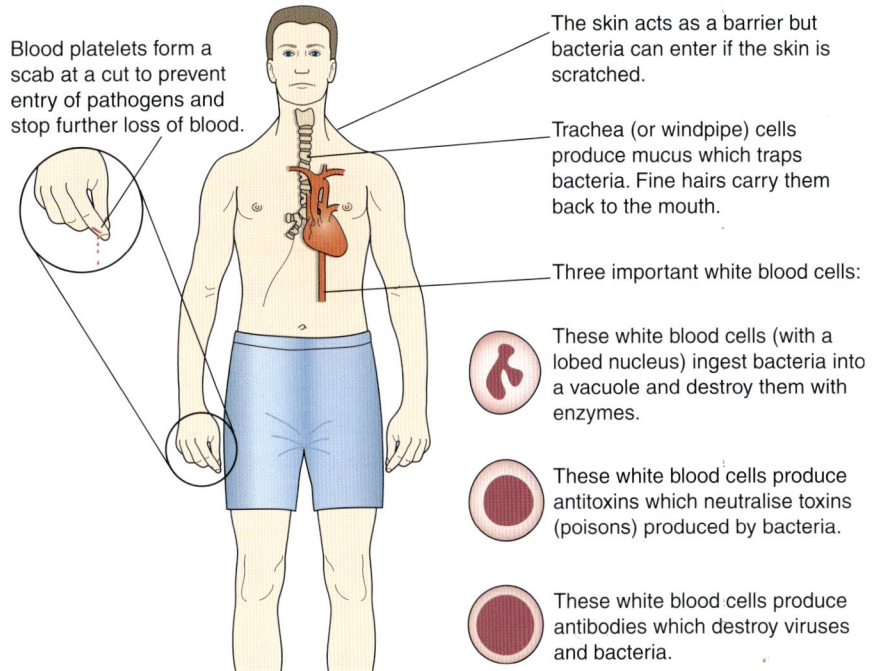

Blood platelets form a scab at a cut to prevent entry of pathogens and stop further loss of blood.

The skin acts as a barrier but bacteria can enter if the skin is scratched.

Trachea (or windpipe) cells produce mucus which traps bacteria. Fine hairs carry them back to the mouth.

Three important white blood cells:

These white blood cells (with a lobed nucleus) ingest bacteria into a vacuole and destroy them with enzymes.

These white blood cells produce antitoxins which neutralise toxins (poisons) produced by bacteria.

These white blood cells produce antibodies which destroy viruses and bacteria.

Figure 2.28 The body's natural defence against disease

All pathogens have unique marker proteins on their surface. These unique marker proteins are called antigens.

When someone catches an infection, their white blood cells respond by producing **antibodies** specific to these antigens. (Specific means that the antibody has a structure which only fits one antigen so there are as many different shaped antibodies as there are different antigens.)

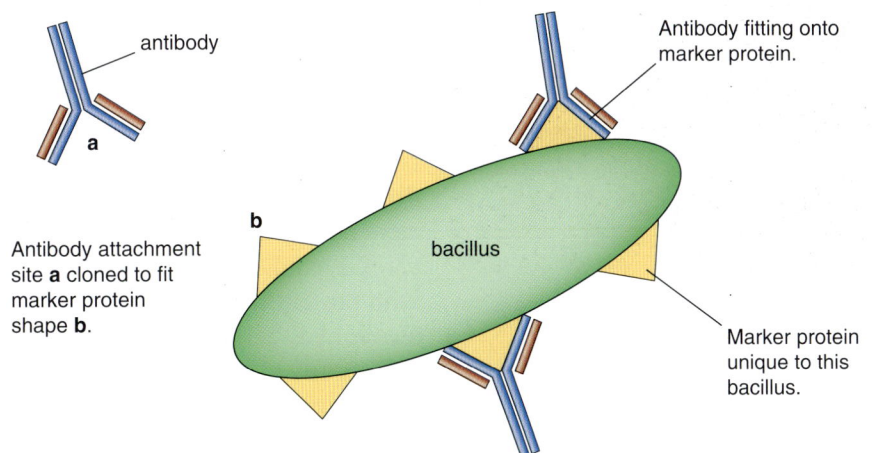

antibody

a

Antibody attachment site **a** cloned to fit marker protein shape **b**.

b

bacillus

Antibody fitting onto marker protein.

Marker protein unique to this bacillus.

Figure 2.29 How an antibody fits onto the protein marker of the pathogen

Antibody production takes a few days. The antibodies cause the pathogens to stick together and these groups of pathogens are then ingested by white cells, as in Figure 2.28. Once the antibodies have been produced, the patient begins to get better. 'Memory cells' are also produced at this point. These 'memory cells' produce antibodies rapidly if the same pathogen invades the body again. This means that a person will not catch the same infection twice.

2.7 Protecting our bodies against pathogens

Our bodies can be protected against some pathogens by **immunisation**. This involves injecting a **vaccine** which contains weak, inactive or dead forms of a pathogen. The vaccine stimulates the white blood cells to produce antibodies specific to the pathogen. 'Memory cells' are also made which make the person immune to further infection by the same pathogen.

> A **vaccine** is a solution which contains weak, inactive or dead pathogens.

Vaccination can also be used to protect children against pathogenic viruses such as measles, mumps and rubella (MMR). When viral vaccines are prepared, the viral genetic material is removed and only the viral coat is used in the vaccine (see Figure 2.25). Following vaccination, antibodies and 'memory cells' are produced. These give the child lasting protection.

Antibiotics

Antibiotics help to cure bacterial infections by killing bacteria inside the body. Different antibiotics attack bacteria in different ways. The most common antibiotic is penicillin. This makes the bacterial cell wall porous so that the bacteria burst. Other antibiotics prevent the bacteria from reproducing and some interfere with bacterial enzymes.

> **Antibiotics** are medicines which kill bacteria inside the body which would otherwise cause further illness.

MRSA is an example of a strain of a common bacterium which has undergone a mutation. The mutation has changed the bacterial chemistry so the MRSA strain is resistant to most antibiotics. Overuse of antibiotics and natural selection of the resistant bacteria has resulted in MRSA becoming more common. Figure 2.31 explains this increase in resistant bacteria. If a patient with an infection is given an antibiotic, the normal bacteria are killed but any mutant bacteria may survive and divide, producing a large number of antibiotic-resistant bacteria. These can be passed on to other people. The best defence against MRSA is careful hand-washing. This has led to a major campaign in all hospitals with posters reminding staff and visitors to wash their hands or use alcohol gel before entering wards or touching patients. Hospital hygiene has improved but this is an area which needs constant attention to detail as bacteria are easily spread. Patients who become infected with MRSA are either isolated or barrier nursed. This means that the doctor or nurse wears a disposable apron and gloves whenever they touch the patient and throws these away afterwards. They must also wash their hands and use alcohol gel before visiting the next patient. This procedure has reduced hospital MRSA infections considerably.

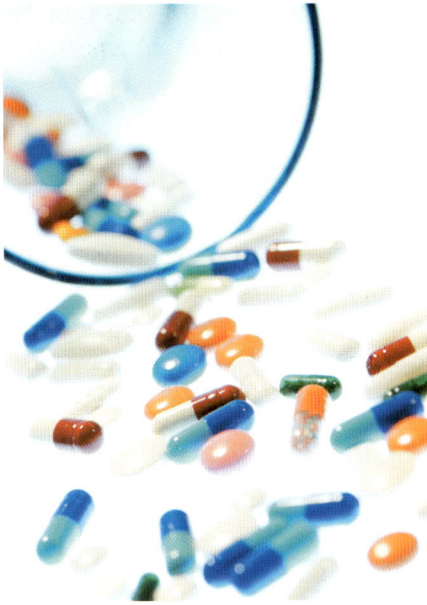

Figure 2.30 There are many different antibiotics

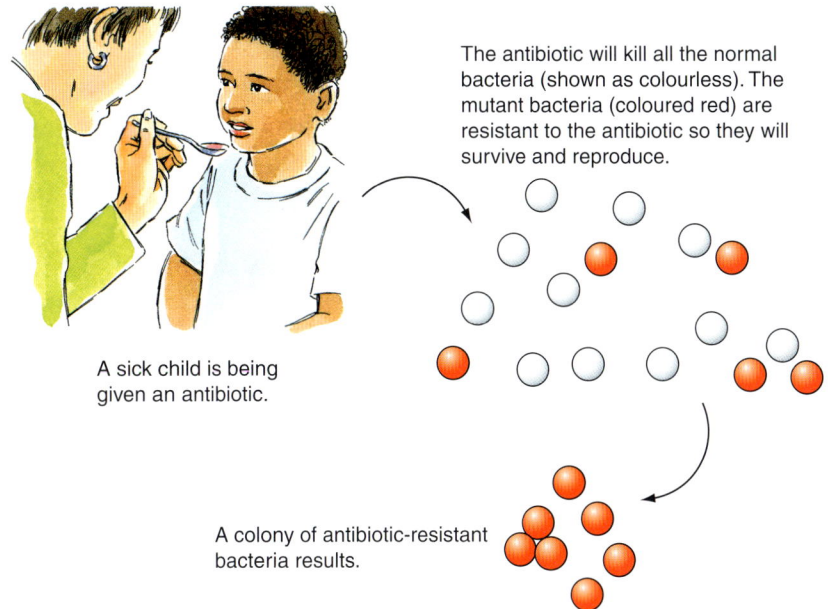

The antibiotic will kill all the normal bacteria (shown as colourless). The mutant bacteria (coloured red) are resistant to the antibiotic so they will survive and reproduce.

A sick child is being given an antibiotic.

A colony of antibiotic-resistant bacteria results.

Figure 2.31 Development of a strain of bacteria resistant to antibiotics

Viruses

Unlike bacteria, viruses cannot be destroyed using antibiotics. Viruses live and replicate inside cells. It is therefore difficult to develop drugs which can enter cells and stop the virus replicating without damaging our body cells. We have to rely on our immune system to fight the virus.

Many people are worried about 'bird flu' (or more correctly, avian influenza). As you can see in Figure 2.25, viruses have a very simple structure. If the DNA from a bird flu virus were to get inside a human flu virus, the new mutated virus could spread through the human population very quickly. This is what happened in the mild flu epidemics of 1957 and 1968. The serious flu epidemic in 1918 was a bird flu virus that adapted to humans. People who rear chickens for meat and eggs have close contact with the birds and are therefore more likely to come into contact with the virus.

The virus is mostly in Asia but any cases in birds in other countries are closely monitored. Avian influenza could spread around the world as ducks and geese migrate to winter feeding grounds. Thousands of wild geese migrate to wetlands in Britain and other parts of Europe every year. The virus is excreted in the droppings of infected birds, so if a few carry the virus it could pass to domestic geese, ducks or chickens. The route from domestic poultry to humans is possible although the evidence is that the virus does not spread easily to humans from birds. However, AIDS is an example of a disease which has probably spread to humans from animals. Any human cases are monitored very carefully and so far no human to human transmission has been confirmed.

> A **vaccination programme** is a schedule of injections given to a child to prevent common childhood illnesses.

The HERD effect

In trying to prevent the spread of a disease in the younger population, the aim is to vaccinate 95% of all children in the UK. If this is achieved the pathogen cannot be transmitted because there are so few unprotected children. Vaccinating most of the 'herd' protects the others.

㉒ The vaccination programme explained in the bullet points on the right has been successful in preventing the spread of infectious disease in the UK. The conditions in a developing country are very different.

Use the bullet points on the right to explain why it is difficult to prevent the spread of an infectious disease in a developing country, such as Uganda.

Vaccination programmes

In developed countries deaths from infectious diseases are rare as a result of the **vaccination programmes**. Vaccination has proved very successful in the UK for the following reasons:

- there is a schedule for vaccination of babies and children up to 15 years of age;
- the majority of children are vaccinated;
- parents receive computer printed reminders for booster injections;
- there are good supplies of safe vaccine paid for by the government;
- there are refrigerated lorries to transport the vaccine;
- there are sufficient local community nurses to carry out the programme;
- most children are healthy and well fed so their immune systems respond by producing antibodies and memory cells.

Summary

✓ **A balanced diet** should contain different food groups. It must also provide essential amino acids and fatty acids, vitamins, minerals and the correct amount of energy.

✓ A diet that lacks any of the above items leads to poor health.

✓ If you take in a larger amount of energy than you use, the excess is stored as fat.

✓ High levels of **cholesterol** in the blood increase the risk of heart disease.

✓ Cholesterol is transported in the blood by lipoproteins. HDLs are 'good' because they remove cholesterol from the blood. LDLs are 'bad' because they cause fatty deposits in blood vessels.

✓ **Metabolic rate** is the rate at which the body uses food and energy.

✓ Exercise raises the metabolic rate.

✓ People who exercise regularly are usually fitter than people who take little exercise.

✓ In developed countries, arthritis, diabetes, heart disease and high blood pressure are linked to obesity.

✓ In developing countries, most diseases are related to food shortage and contaminated water.

✓ Too much salt in your diet is a risk factor for heart disease.

✓ Processed ready meals often contain high salt levels.

✓ **Pathogens** are microorganisms which cause disease.

✓ The body's protection against disease includes the skin, mucous membranes and white blood cells.

✓ **Antibiotics** are medicines which kill pathogenic bacteria in the body.

✓ **Bacteria** reproduce rapidly and can undergo genetic mutations. MRSA is a mutant strain which is resistant to many antibiotics.

✓ Hospitals can control the spread of bacterial infections by strict hygiene control.

✓ In the nineteenth century, Semmelweis observed that hand-washing reduced the spread of infection.

✓ **Immunisation programmes** have been very effective against infectious disease.

✓ **Vaccines** stimulate the white blood cells to produce antibodies specific to the infecting pathogen.

EXAMQUESTIONS

❶ Which of the following are causes of obesity?
A lack of exercise
B drinking too much water
C overeating
D eating seven servings of green vegetables each day   (*2 marks*)

❷ Which of the following could result from a diet with a large amount of red meat?
A excess of calcium
B saturated fats in the blood
C anaemia
D a high level of cholesterol in the blood
                                              (*2 marks*)

In questions 3–5, only one of the alternative answers is correct.

❸ People who exercise are generally fitter than those who don't because
A they have a low metabolic rate
B they have small muscles
C their heart muscles are stronger
D they have high blood pressure.   (*1 mark*)

❹ Antibodies prevent infection by
A causing bacteria to stick together
B sterilising the skin
C destroying bacterial cell walls
D producing sticky mucus.   (*1 mark*)

❺ This question is about MRSA. Which statement is *incorrect*?
A MRSA is resistant to many antibiotics.

B MRSA has developed resistance as a result of mutation.
C MRSA can be stopped from spreading by good hygiene.
D A person infected with MRSA cannot be cured.   (*1 mark*)

❻ A person eats some food contaminated with food poisoning bacteria.
a) Which part of the body will these bacteria infect?   (*1 mark*)
b) The person is ill for two days and then begins to feel better. Draw a diagram of the white blood cells that produce antibodies.
                                              (*1 mark*)
c) Draw a diagram to show how an antibody is specific to one pathogen. Label this diagram.   (*3 marks*)
d) How do memory cells prevent a person catching an infection a second time?
                                              (*2 marks*)

❼ a) How should doctors and nurses try to reduce infections passing from one patient to another?   (*2 marks*)
b) Give two ways in which antibiotics can kill bacterial cells.   (*2 marks*)
c) Explain how a bacterium can become resistant to an antibiotic.   (*3 marks*)

Chapter 3
What determines where organisms live?

At the end of this chapter you should:

✓ be able to identify special adaptive features of animals;
✓ appreciate how adaptations allow an animal to survive in hostile environments;
✓ recognise the adaptations of plants for different environments;
✓ understand competition among plants and animals;
✓ understand the consequences of an increasing human population on the depletion of raw materials;

✓ be aware of the reasons for an increasing area of land use by humans;
✓ be aware of an increasing production of waste and pollutants by humans;
✓ be able to evaluate scientific evidence and separate this from non-scientific opinion;
✓ appreciate the factors involved in decisions about sustainability.

Figure 3.1 Commercial logging in an Indonesian rainforest causes widespread damage to the ecosystem

3.1 What determines where a particular species lives?

At KS3 you investigated food chains, food webs and 'who ate whom'.

In this chapter we will look, in greater depth, at the adaptations of organisms to the specific environments in which they live and the factors which control population size.

Animals are said to be **adapted** to their surroundings when they are able to survive and reproduce successfully. In this case, the environment provides a supply of food for growth and a nest site or place to raise their young.

If you were going on a field trip to the Arctic, what clothing would you take? You would probably take a thick fleece and windproof jacket. Animals that live in the Arctic must also be able to withstand the cold conditions.

The photograph in Figure 3.2 shows the features or adaptations which enable a polar bear to survive in the Arctic. The polar bear has:
- a small head and ears;
- a compact body shape which is also streamlined for swimming;
- a thick layer of fur – this traps air, which is a good insulator;
- a thick layer of fat (up to 11 cm thick) which insulates against the cold and also acts as a food reserve during the Arctic winter when the bear is in a deep sleep;
- white fur which camouflages the bear, enabling it to get closer to prey when hunting;
- white fur to reduce heat radiated from the body.

Compare these features of the polar bear with a camel.

A camel lives in hot arid conditions. It needs to prevent overheating and is adapted to this by having:
- long legs and neck giving a large surface area for heat loss;
- thin hair on top of the body to allow heat loss;
- no hair on the underside of its body making heat loss easier;
- little body fat so heat is easily lost from the skin capillaries;
- a fatty hump which can act as a food reserve when desert food is scarce;
- camouflage colouring;
- two rows of eyelashes;
- nostrils which close for protection during sandstorms.

Many plants are also adapted to live in harsh environments with features to stop them losing too much water. The house leek (Figure 3.4) lives on rocky outcrops where rainfall varies during the year with none at all

Figure 3.2 Polar bears are adapted to survive in cold conditions

Figure 3.3 The Arabian camel is adapted to survive in hot, dry conditions

An **arid** region is hot and dry. This results in sparse vegetation which is usually adapted to reduce water loss and attack by herbivores.

Figure 3.4 A house leek showing the fleshy leaves with a waxy coating

during the summer. Similar conditions are found when they are grown on roofs or in rock gardens. Water loss is reduced by having:

- a short stem;
- fleshy green leaves which store water but which dry up at the end of the year;
- a waxy, shiny outer covering to the leaves;
- long roots which penetrate deep into the soil in the rock crevices.

From these examples you can see that plants and animals can survive in different conditions because of the special adaptations which suit their environment. Look carefully at the photographs of two different foxes and answer the questions below.

Figure 3.5 Arctic fox

Figure 3.6 Fennec fox

❶ a) Which animal is adapted to retain body heat?
  b) Explain how its body features help to retain heat.

❷ a) How do the fur colours relate to their environment?
  b) The Arctic fox moults its coat in the spring. What colour would you expect the summer coat to be?

❸ What is the purpose of the thick layer of fat on the Arctic fox?

❹ The Arctic fox has short legs covered with thick fur. What would you expect the legs of the fennec fox to look like?

❺ Why do the two foxes, which both listen for prey, have such different sized ears?

3.2

What determines the number of organisms in an environment?

Figure 3.7 In this hedgerow the primroses have flowered before the trees have opened their leaf buds

Plants compete for light to photosynthesise and for soil space from which to absorb water and take in minerals. This interaction of organisms (plants or animals) trying to obtain the same food or occupy the same territory is called **competition**.

Consider a hedgerow beside a road. Plants such as primroses, like those in Figure 3.7, flower early in the year to avoid competition for light. They also produce leaves, flowers and seeds before the tree leaves open and put them in shade.

Animal populations are also regulated by competition among members of the same species and competition between different species. Some animals and birds compete for a territory of sufficient size to feed their offspring. Although people love to hear blackbirds and robins singing, their songs are actually 'war cries' to keep other males off their territory. These birds sing from different high points around the area to mark the boundary. Can you name any animals that have a territory? How do they mark it?

Figure 3.8 A gannet colony. The nests are evenly spaced and just out of reach of the bird at the next nest

Gannets are sea birds that catch fish by diving head-first into the water. They live and breed on remote rocks or cliffs. Just imagine the noise in the colony when all the birds are competing for mates and nesting sites.

Notice in Figure 3.8 that the nests are placed 'pecking distance' apart. You can see why when you look at the sharp, pointed beak of the gannet! The size of a gannet population changes with food availability. If more fish are available, more young gannets are raised, but this increases the competition for nest sites in future years. Other limiting factors within the gannet colony are **predators**, such as gulls, who brave the gannets' spear-like beaks to steal eggs or young.

In general, the number of organisms in an environment is strongly influenced by food supply. Numbers will increase if there is plenty of food and decrease if food is in short supply. From the examples given, you can see that population size is also determined by competition among members of the same species, and by competition with other species for the same nutrients or space. This applies to both plants and animals.

Figure 3.9 Gannets have very pointed beaks

A **predator** is an animal which catches and eats another animal. The animal caught is called the **prey**.

3.3 How do humans affect the environment?

There are 6 billion people in the world today and the number is increasing.

How can the requirements of this increasing human population be met without damage to the environment? People in the UK expect to have a house or flat to live in and access by road to supermarkets and other shops. In their homes they want electrical goods such as a fridge, freezer

An **ecosystem** is made up of the plants and animals living and surviving in one place and interacting with the surrounding non-living environment.

A **biological control species** is an organism that limits the numbers of another species that is considered a pest. (For example, ladybirds eat greenfly – the ladybird is the control species and the greenfly is the pest.)

Figure 3.10 A road widening scheme carves through woodland, dividing the ecosystem

Activity – Campaign poster

A local council is proposing to build a new road through the middle of an area of ancient woodland which contains rare plant and butterfly species and a small population of deer. It is also a popular area for families at weekends. Prepare a campaign poster outlining the reasons for re-routing the road around the wood. Use illustrations to add interest.

and washing machine, and many have a computer and media system. Can all these be provided without impact on the environment? To answer this question we will examine the issues of a) land use; b) raw materials; c) waste, and d) waste-water treatment.

a) Does an increasing population reduce the amount of land available for animals and plants?

An increasing human population needs more land for:
- building homes, industrial estates, motorways and retail parks;
- quarrying and mining of raw materials for buildings and roads;
- intensive farming;
- waste disposal.

Currently between 10 and 20% of the UK's native species are threatened with extinction by expanding towns, motorways and intensive agriculture. Sites of Special Scientific Interest (SSSIs) have been established in some areas to support a rare species or group of species. Such species need special protection and the ecosystem around them needs to be large enough to remain balanced and undisturbed. Size is very important when setting up an SSSI. However, there can often be problems if a rare species is located where a motorway is planned!

A balanced **ecosystem** will change slowly over time. This means that the number of plants and animals and the range of species remain similar from year to year. Animals and plants need a certain area to sustain a breeding population. Dividing up an area of land, by putting a motorway through it, may result in the separate parts suddenly becoming too small to feed populations of larger animals such as deer or badgers.

Having examined the consequences of making an ecosystem smaller, consider the opposite case of making it larger. Removing hedges and ditches to make bigger fields removes the shelter, feeding and breeding sites for animals, including **biological control species**, such as predatory beetles, insectivorous birds and hedgehogs. These are the carnivores in the food web. By eating pests such as greenfly and caterpillars, these predators can naturally keep pest numbers low. Removing hedges and ditches upsets the balance of the ecosystem, disrupting the food webs and allowing pests to reproduce and cause further crop damage.

b) Does an increasing population have an impact on raw materials?

Important raw materials, such as building stone, metal ores and fossil fuels, are finite resources. Their increased use means they will run out more quickly and alternatives, such as renewable energy sources and recycled building materials, must be developed. Trees take many years to grow to a size where the wood can be used, so regular replanting must take place to prepare for future needs. Our use of raw materials is considered further in Sections 5.1 and 5.6.

Figure 3.11 Landfill sites attract scavengers, such as gulls

6 Imagine an incinerator is being built near you and the campaign against its construction is being supported by your local newspaper.
a) Do you think the newspaper would report scientific evidence for and against the incinerator or people's feelings?
b) Do newspapers generally print scientific, factual information or eyecatching headlines?
c) Do you think a newspaper is always a reliable source of information, in such cases?
d) i) Would you expect to see a report with data in your local paper?
   ii) If the data were produced by the company wanting to build the incinerator do you think they would be biased?
e) Technology and environmental regulations are regularly improved – do you think the newspaper information would be up-to-date?

c) Can the volume of waste deposited in landfill sites be reduced?

Low-lying areas of land and old quarry pits are often used as landfill sites, being filled with waste which is then compressed. Although these sites are well regulated there are still environmental problems associated with them. Sites can be unsightly. Pathogens can be transported by birds such as gulls, which feed on the waste, and toxic chemicals can be washed into waterways after heavy rainfall.

Landfill sites are not an ideal solution to the disposal of our rubbish. Many sites around the UK are now almost full and, in some cases, no local alternative site is available.

Waste disposal is a complex problem. New solutions to the increasing problem of waste disposal need to be found.

Currently:
- the UK produces over 100 million tonnes of waste per year;
- the increasing population of the UK and higher standards of living mean that waste production is expected to rise;
- a large percentage of rubbish is disposed of in landfill sites;
- a considerable proportion of the material placed in landfills could be recycled to save raw materials;
- the government initiative 'Waste Strategy 2000' plans to reduce landfill by 45% by 2010 and 67% by 2015.

So what can be done with all our rubbish? Alternatives to landfill include: recycling, incineration and composting. All of these alternatives require processing plants, which cost money to build although some of the costs could be met by the sale of materials.

Recycling
This can involve the mechanical sorting of mixed rubbish (for example, pulling iron / steel out using a giant electromagnet) or sorting by each household. See Section 3.6.

Incineration
This process further reduces the volume of rubbish placed in landfill sites. The ash produced can be used for road building, thus saving natural stone. All new incinerators are also EfW plants (Energy from Waste). This means that the heat produced from burning the rubbish is used to make electricity. Proposals for building new incinerators frequently encounter opposition from local residents who fear the risk of air pollution. Modern incinerators are, however, highly sophisticated industrial plants where all emissions are released at safe levels and monitored by a computer 24 hours per day.

Composting
This is used to reduce the volume of **biodegradable** materials. After treatment (to kill pathogens) the compost can be used or sold.

Eutrophication is the nutrient enrichment of a river by the addition of nitrates, phosphates or ammonium compounds. As a result, algae reproduce and the water soon becomes green. Rooted water-plants grow rapidly and this slows down the flow of water. Algae have a short life cycle. Death of the algae is followed by an increase in aerobic decomposer bacteria which break them down. These aerobic bacteria use up oxygen and this deoxygenates the water.

Biodegradable material is usually of plant or animal origin. It can be decomposed by bacteria or fungi. (Paper, cotton and wool waste are all biodegradable.)

d) What is the impact of an increasing population on waste-water treatment?

Before the Industrial Revolution people lived in small, scattered rural communities. Waste water was emptied into ditches and the biodegradable contents decomposed. Now, towns have much larger populations. If sewage were emptied into our rivers it would create an awful smell and fish and other wildlife would be poisoned. The treatment of waste water and sewage has improved to meet the demands of an increasing population. Today only treated sewage passes into rivers or out to sea and it must not contain pathogens or toxins (poisons) or cause **eutrophication**.

Many pharmaceutical companies and power stations employ a biological solution to overcome waste-water problems. They filter waste water through reed beds. The reeds use the nutrients in the waste material to grow and at the same time remove toxins from the water. Reed beds provide an interesting habitat for wildlife. Water run-off from motorways, and some airports such as Heathrow airport, is treated by flowing through such reed beds.

Section 3.3 started with the question 'How do humans affect the environment?' After examining four different issues, we can now make some positive suggestions. As a society, we must:

- avoid changing balanced ecosystems as this can have serious consequences and damaging long term effects;
- increase recycling and the re-use of materials such as metals, plastics, glass and paper. This will save raw materials;
- reuse **brown field sites** in town for new housing. This will save farmland and new roads will not be needed;
- use biological methods where possible to treat waste water and biodegradable waste, as the end products are harmless to the environment.

Figure 3.12 Reed bed filtration – an opportunity for a new ecosystem

7 An increasing population leads to increased transport. List three ways in which this can affect ecosystems.

8 The organisms shown in Figure 3.13 eat pests which damage plants.

a) What is the name given to this useful group of insects?
b) How have numbers of the biocontrol species been reduced?

Figure 3.13

3.4 Are the problems associated with an increasing population restricted to developed countries?

The term **brown field site** is used to describe an area in town where old properties have been demolished. A **green field site** is farmland or green belt land surrounding a town where there have been no buildings before.

Large-scale deforestation of tropical rainforests throughout the world has been, and still is, widely publicised. The Amazon rainforest is a major source of global oxygen and a large number of pharmaceutical drugs have been developed from rainforest plants. We therefore must take steps to ensure the survival of this resource. It also shows that the problems associated with an increasing worldwide population are not restricted to developed countries.

The ecosystem of a rainforest is dependent on nutrient recycling. Dead leaves are decomposed by bacteria and fungi, and the minerals are released into the soil. 'Slash and burn' also releases minerals into the soil which can support the arable crops of a small population for a few years. As minerals in the soil are removed and not replaced, the crop yields fall and the village population moves, abandoning the clearing. The natural vegetation will regenerate from seeds in the surrounding forest, but this takes time. Very different effects result from commercial logging (compare Figure 3.14 with the photograph at the opening of this chapter).

Deforestation provides:
- timber that is sold for money;
- timber for use as a building material;
- land for cattle ranching or growing crops to generate income;
- land to build roads which connect major cities.

Figure 3.14 A small tribal clearing in the rainforest

9 Why do small clearings cause less damage than large-scale logging?

10 a) Who do you think provides the evidence for the re-growth of vegetation in small cleared areas?
b) What type of evidence would you need to be convinced that regeneration takes place? Briefly suggest a project to investigate this.
c) What is it about the soil in a rainforest that limits its value as agricultural land?
d) Why does soil erosion occur in deforested areas?
e) i)   What did scientists from the Massachusetts Institute of Technology discover?
   ii) Do you think their evidence can be reliably linked to flood damage reported in vast areas cleared of forest?
f) List what you see as the harmful long term consequences of extensive deforestation of rainforest. Will the effects be local, global or both?
g) Using your answer to question f), suggest a management strategy that would result in sustainable development of the rainforest. (See Section 3.6 for an explanation of sustainability.)

Large-scale logging results in:
- vast cleared areas;
- removal of the tree canopy, which breaks heavy rainfall;
- compaction of the soil and root damage by heavy machinery;
- poor soil drainage and no aeration to the roots of small plants due to heavy machinery;
- soil erosion as water from heavy rainfall flows across the land;
- flooding when soil washes into streams and rivers and causes a blockage.

Scientists from the Massachusetts Institute of Technology examined 75 years of rain-gauge records. The records showed a big increase in rainfall over large deforested areas. The scientists also examined current satellite data of cloud cover. Satellite pictures showed twice as much low-level rain-bearing cloud over large deforested areas, compared with forested areas. Small deforested areas do not give the temperature differences needed for the formation of clouds.

Other effects of commercial deforestation include:
- isolation of animal populations which will not cross large open areas where forest has been cleared;
- reduction in the sources of food for both human and animal populations.

Approximately 160 000 square kilometres of deforested land have already been abandoned because the thin, acid soil cannot support crops or cattle. These areas are slowly becoming deserts. The Brazilian government uses satellite surveillance technology to help prevent illegal logging. It faces the dilemma of trying to improve the lives of the increasing Brazilian population while at the same time preserving the rainforest.

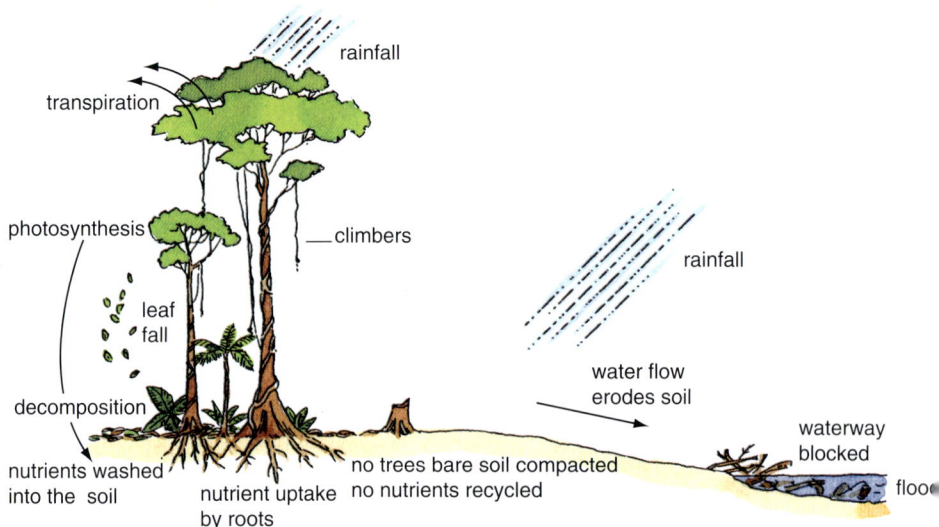

Figure 3.15 The rainforest cycle and biological consequences of disturbance

More carbon dioxide produced by respiration and decomposition.

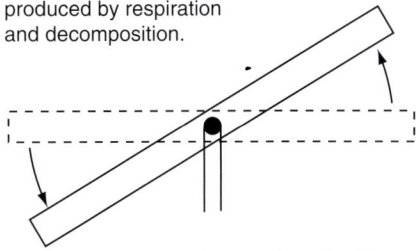

Less carbon dioxide used for photosynthesis.

Figure 3.16 Deforestation results in an increase of carbon dioxide in the atmosphere

A rainforest appears lush but it is still a delicately balanced ecosystem (see Section 3.3). In an undisturbed ecosystem the carbon dioxide uptake for photosynthesis is balanced by that produced by respiration and decomposition. Logging results in the release of higher levels of carbon dioxide by:

- increased activity of soil microorganisms, as leaves and branches from felled trees decompose;
- burning of waste wood in fires.

Trees convert or 'lock-up' carbon from carbon dioxide into compounds such as cellulose during their period of growth. The ecosystem balance is upset when all the carbon that was 'locked-up' as cellulose is suddenly released in one fire. Slow plant regeneration means that there will be little photosynthesis as the ecosystem recovers and so the carbon dioxide remains in the atmosphere. Destruction of rainforest can therefore play a significant part in rising carbon dioxide levels.

3.5 Global warming

There is an interesting link between deforestation, cattle ranching and global warming. Deforestation provides land for cattle ranching. Cattle are ruminants. A large part of their stomach, called the rumen, acts as a fermentation chamber.

Anaerobic bacteria live where there is no oxygen. Methane (CH_4) is produced if carbon compounds are broken down when there is no oxygen present.

There are anaerobic bacteria inside the rumen that produce enzymes which break down cellulose and provide sugars for energy. One of the waste products of this process is methane, which is released into the atmosphere. Unfortunately methane is a potent greenhouse gas.

The anaerobic bacteria in the rumen of cattle are not the only methane producers. Methane is also produced in paddy fields by anaerobic bacteria that decompose dead vegetation in the waterlogged conditions. As the population in Asia increases, more rice is required and so more land is flooded. This leads to an increase in the volume of methane released. Methane is also produced by the anaerobic decay of materials in sewage and household refuse. In some countries, this methane is collected, stored and used as fuel. In the UK the methane is burned and used to control the temperature of the anaerobic sewage digester tank.

Other aspects of global warming and the greenhouse effect are covered in Section 6.4.

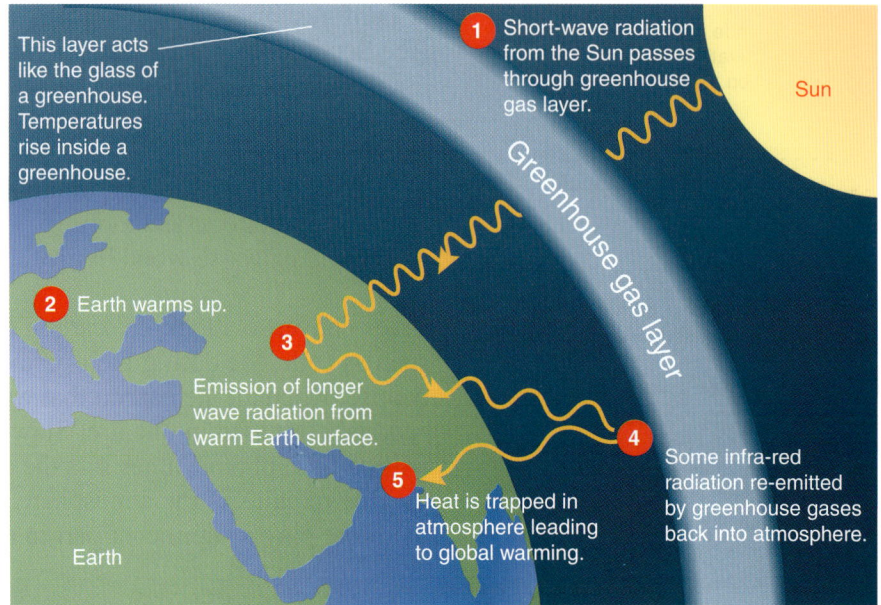

Figure 3.17 The greenhouse effect

How science works – the use of models for complex problems

A **scientific model** is a hypothesis which can be represented as a mathematical equation or a computer program. Models are valuable when there are many different factors which affect the outcome of an experiment. The accuracy of the model can be tested by applying it to data that has been collected in the past. If the model fits with past data, it can be used to simulate what might happen in the future. Weather forecasts are created using scientific models which incorporate many factors – if one factor changes unexpectedly the forecast may be wrong!

We have already seen that ever increasing amounts of carbon dioxide and methane are being produced. Most scientists believe that these so-called 'greenhouse gases' are largely responsible for global warming. Other scientists believe that global warming has taken place before and that we are in a new warming cycle which has nothing to do with these gases. When 'sceptics' challenge scientific ideas further research is carried out, further **scientific models** are produced and tested. With more detailed and reliable information scientists get closer to the truth. The 'global warming' debate takes place not only among scientists but also among politicians, environmentalists and humanitarians. Figure 3.18 shows the different bias which each group brings to the debate.

Figure 3.18 The global warming debate

Should we be concerned about the greenhouse effect and global warming?

By analysing records over many years, scientists have made the following observations:

- the cold winter period in the Arctic has become shorter and the ice caps are melting;
- sea levels have risen between 10 cm and 20 cm in the last hundred years as the snow cover in the Northern Hemisphere has melted;
- graphs of temperature and carbon dioxide show increases which follow similar trends;
- in recent years there has been unusually heavy rainfall and flooding in the UK and Northern Europe;
- droughts and crop failures in the Sudan and other regions have increased in recent years.

All these observations fit the predictions made by scientists and meteorologists (weather forecasters) based on their current scientific models.

Although there has been a great deal of debate about global warming, it now appears that scientists have underestimated the rate of rise in world (global) temperatures.

A rise in temperature of only a few degrees Celsius is enough to change the climate. This could mean a change in direction of rain-bearing winds, resulting in heavier rainfall in some parts of the world and severe drought elsewhere. If the melting of ice continues there will be a further rise in sea level and areas of Holland, Eastern England and Bangladesh which are just above sea level could be flooded.

Having collected and analysed evidence from measured data and scientific models we can start to form conclusions. All governments should be concerned about global warming and they should take action to reduce their emissions of carbon dioxide and methane. Individuals should also play their part. This will be explored in the next section.

3.6 What can be done to reduce human impact on the environment?

Individuals, local government and industry can all make changes, based on scientific evidence, to reduce the negative human impact on the environment. People who recycle are acting responsibly to conserve natural resources. We *all* have a part to play in meeting the targets of local and national schemes to reduce waste. Local authorities are making recycling easier for us. Not only are there bottle banks and facilities for recycling aluminium cans, but many towns now also have

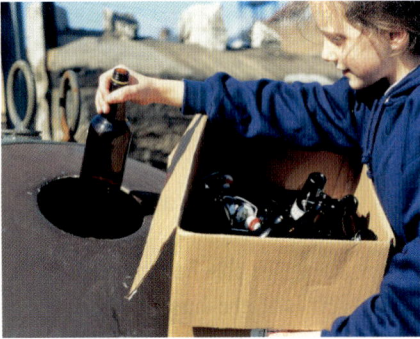

Figure 3.19 More and more young people are actively recycling and trying to reduce waste. Is there a point to this?

kerbside collections for newspapers, bottles and cans. Countries such as Germany have been doing this longer than the UK and have significantly reduced waste.

Activity – 'Waste local plan'

❶ If your town has kerbside collections, find out which materials are collected separately.

❷ Make a list of common products that can be produced from recycled materials.
You may find www.recycledproducts.org.uk useful for your research.

The energy-efficient home: saving fuel saves fuel costs

Making our homes and offices more energy efficient by installing loft insulation, cavity wall insulation and double glazing reduces the amount of fuel needed for heating (see Chapter 9).

These changes are beneficial because:
• fossil fuels will last longer;
• greenhouse gas emissions will be reduced;
• gas, oil or electricity bills will be reduced.

❶❶ Suggest other ways that you could reduce energy used a) in your home and b) within your school. Visit www.est.org.uk before answering this question.

Why walk or cycle?

Walking or cycling instead of driving:
• saves fuel costs of private vehicles;
• reduces air pollution;
• saves petrol and oil, which are non-renewable resources.

Public transport, which can carry many people, provides similar benefits to the environment.

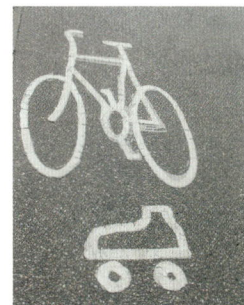

Figure 3.20 Where is your nearest cycle route?

Activity – What can you do to make a better future?

This is a group activity. Log onto www.sustainable-development.gov.uk for some ideas.

Alternatively, you can log onto the RSPB website www.rspb.org.uk. Type 'green living' into the search box.

1. Planning

   In a small group, develop an environmental plan that involves avoiding waste or increasing recycling. Try to think of a new idea that you could put into action in either your school or the community in which you live. For example, a local charity might have a fund-raising recycling project for which you could provide support.

2. Communicating

   Create a poster, webpage or PowerPoint presentation which outlines your plan to others in your class. Remember to explain the importance of each suggestion in your plan. For example, recycling aluminium cans saves the raw material and the energy used to refine aluminium ore.

3. Action

   Obtain permission to carry out your plans and put them into action! Remember, you are more likely to keep up simple tasks, such as collecting cans for recycling at break or organising a petition for a can bank, than more complicated ideas.

Sustainable development is 'development to improve the quality of life for people living now without reducing the ability of future generations to meet their own needs' (United Nations definition).

Figure 3.21 The World Summit on Sustainable Development, Johannesburg 2002

Do governments consider the ethical issues when planning for future development?

Even within the UK there are issues of poverty, pollution and health care that need to be addressed. The UK government has already produced a **sustainable development** strategy. This means planning to live 'within the means of the environment' so that natural resources will still be available for future generations. One part of the strategy seeks to control fishing, to allow fish to grow to reproductive size and breed before they are caught. This should allow the UK fish population to increase and prevent fishing to extinction. Another part of the strategy encourages the redevelopment of brown field sites (previously developed land in towns) rather than developing and reducing the area of farmland or parkland, and building houses which ordinary people can afford.

You can log onto the UK's government website at www.sustainable-development.gov.uk for more information.

More details can be found at www.uyseg.org. Double click on the Sun to enter the site and then 'Chemical Industry Education Centre'. Click 'webcentre', 'secondary' and then 'sustain-ed'.

A brief history of the world efforts for sustainability

The first United Nations Conference, to consider the impact of human activity on the environment, took place in 1972. In 1992 more than 100 countries met for the first international Earth Summit. This summit produced a major plan for sustainable development, reducing poverty and giving people access to resources to support themselves. The plan called upon governments in all UN countries to lead the way in reducing pollution, emissions and the use of natural resources. In 2002 the Johannesburg Summit reviewed the little progress that had been made in the previous 10 years and focused on reducing poverty, and increasing access to safe drinking water and sanitation.

Activity – Easter Island

Figure 3.22 Easter Island was once a wooded island

Easter Island is known for its amazing statues, up to 10 metres high, which were carved in stone about 1200 years ago. Today, there are no trees on the island. But archaeological investigations have shown that the island was once heavily wooded when the first inhabitants arrived by canoe 1300 years ago. Easter Island palm trees were ideal for building houses and hollowing out for canoes which were used for catching seafood, especially porpoises. The statues provide evidence of a complex culture, because engineering skills must have been used in transporting the enormous statues to sites all around the island. This was done by rolling the statues on logs and pulling them with ropes made from climbing plants. The population on Easter Island increased rapidly but this resulted in deforestation followed by wind erosion of the soil, poor crop yields and the drying-up of water springs. Without palms to construct canoes, there were no porpoises to eat. Once the inhabitants of the island had eaten all the sea birds and shell fish, starvation, anarchy and cannibalism set in. This sad story was uncovered from the pollen record and bone deposits in rubbish sites on the island.

Is this a picture in miniature of our effect on other ecosystems? Year by year, there is evidence of similar damage to environments and

tribes in the Brazilian rainforests, Africa and Asia. Is it time to act now for sustainability before it is too late?

❶ Copy and complete the following table, comparing the disaster on Easter Island with tribal use and commercial logging of the Amazon rainforest.

	Easter Island	Tribal use of rainforest	Commercial logging of rainforest
Timber for	• Building homes • Building boats/ canoes • Rollers to transport statues • Fuel		
Main food supply		• Bush animals • Fruits, berries and roots • Grubs	Food is delivered to logging camps and results in food waste and packaging waste
Soil erosion	Yes – as a result of tree removal		
Animal populations	Reduced by over-fishing and habitat destruction	Little damage – small areas used in which plants regenerate quickly and animals spread back from nearby areas	
Problems faced by humans	Starvation as a result of over-fishing and habitat destruction		Conflict between displaced tribes and loggers
Can the environment be restored?			

Table 3.1

Living organisms as indicators of pollution

A **biological indicator** (**bioindicator**) is a plant or animal which can indicate the level of a particular chemical in its habitat. In the examples in this section sensitive lichens indicate sulfur dioxide levels in the air and invertebrates indicate oxygen levels in water.

Chemical analysis of water or air for pollutants gives a measure of what is present at the time of sampling, but it gives no information about pollution incidents which may already have been diluted or blown away. But, such incidents may kill certain species, leaving evidence for the biological detective.

Pollution indicator species
Different species of lichens have different sensitivities to sulfur dioxide (Table 3.2). Some are so sensitive that a trace of the gas in the air will kill them. This means that by looking at the lichens still growing, we

can use them as **bioindicators** to find out the prevailing levels of sulfur dioxide pollution in the air.

Information from a lichen survey could help a local authority to monitor changes resulting from increased traffic density or industrial activity. This would enable it to make decisions about locations for new roads or industrial sites.

Lichens are pronounced 'li kens'. They are a mutual association of a fungus and an alga. They occur as crusty patches on rocks, tree trunks and walls. Lichens are living organisms which can be used as **biological indicators** (bioindicators) of pollution as the algal part is rapidly killed by sulfur dioxide in polluted air. Lichens grow very slowly over many years.

Type of lichen	Common orange lichen – quite tolerant of moderate pollution levels	Quite a common lichen but dies quickly if pollution levels rise	A common woodland lichen and an indicator of clean air	Beard lichen – only survives in pure air
Maximum level of sulfur dioxide tolerated	$70 \, \mu g/m^3$	$60 \, \mu g/m^3$	$50 \, \mu g/m^3$	$35 \, \mu g/m^3$

Table 3.2 Common lichens which act as bioindicators of air pollution from sulfur dioxide

Activity – Carrying out a pollution survey

A science class decided to survey the air quality around their town, using lichens as indicators.

Starting hypothesis: The lichens growing will be those able to survive in the level of sulfur dioxide pollution found normally at that location.

Prediction: As sulfur dioxide is produced by combustion of fossil fuels, the levels of the gas will be highest near main roads, residential and industrial areas. Therefore, only tolerant lichen will be found in these areas. The lowest concentrations of sulfur dioxide will be found on the SW side of the town in the old woodland and farmland areas and so more sensitive lichen will be found here.

Procedure

Before starting all the students made a copy of the town map on which they marked industrial areas in blue, coloured the main roads in red and all parks, woodland and playing-fields in green. From the town centre they drew eight lines along the main compass bearings (N, NE, E, SE etc.) and made concentric circles from the town centre 0.5 km apart, up to a 5 km radius. Survey points were chosen where the compass lines and circles crossed. Anyone living near an intersection surveyed that location. More distant or difficult points were visited by a group with their teacher. They were told to look on the nearest walls, trees, or hedges to find lichens that were a close match to the photographs of the indicator species.

On return to the laboratory, the results were plotted onto the map.

Figure 3.23 The class evaluated their results together. Here are some of their comments

The speech bubbles read:

chemical spraying on farmland might make a difference,

the sizes and ages of the trees surveyed were different,

some areas were newly developed and had no lichens at all,

more types of lichens were found than given on the pictures,

it is not easy to separate the variables in this investigation,

not enough results to draw a definite conclusion,

Figure 3.24 Results of the lichen surveys

❶ Which gas are lichens most sensitive to and what produces this gas?

❷ a) What was the direction of the prevailing wind?
b) How could this affect the distribution of sulfur dioxide?

❸ a) Where were the most sensitive lichens found?
b) Did this fit the prediction made?

❹ Students reported finding dead and dying lichens in the hedge beside the road running SE from the old town. What does this suggest?

❺ a) Where were the least sensitive orange-coloured lichens found?
b) Look at the map and give reasons that support your answer.

❻ The churchyard in the town centre had a variety of grey and orange lichens but none was found around the newer church to the east of the town. Suggest why.

❼ What instructions were given to students going to a survey point?

❽ Using the map, write a conclusion, stating what the survey found.

Figure 3.25 Students kick-sampling in shallow water

A **fair test** is one in which only the independent variable being tested has been allowed to affect the dependent variable. All other factors which could affect the outcome have been kept constant.

Bioindicators for water oxygenation

Small animals, particularly invertebrates and pollution-sensitive fish, can be used as bioindicators for water cleanliness. Some years ago, there was great excitement when trout were caught in the river Thames after not being seen there for many years. But trout are not usually used as bioindicators in checking water purity; smaller invertebrates are more useful. They can be caught easily in shallow streams (see Figure 3.25). The normal method is kick-sampling, using a D-shaped net to catch the invertebrates disturbed from the stones and plants. The investigator wades across the stream, shuffling the stones with his / her feet. The net is held downstream beside the feet, in the direction of flow, to catch the specimens. For a **fair test**, the sample is collected by shuffling the same distance and for the same time when each sample is taken. The catch is then emptied into a white tray with a little water in the bottom. Specimens can be sorted using a teaspoon, pipette and paintbrush. Each different species is recorded as rare, frequent or in large numbers. A clean well-oxygenated stream will have a high biodiversity (many different species). A polluted stream will have a low number of species which have adaptations to survive in low oxygen conditions. An example of such species are chironomid midge larvae, which are red because they have haemoglobin to pick up oxygen in poorly oxygenated water.

Figure 3.26 A kick-sampling catch being sorted

clean well oxygenated water   poor quality water, low oxygen level

Mayfly nymph

Fresh water shrimp

Fresh water louse

breathing tube stretches to surface air

Caddis larva

Chironmid midge larva

Stonefly nymph

with case made from small pieces of wood

Caddis larva

with case made from tiny stones

These species indicate good quality water.

red colour (haemoglobin)

Rat-tailed maggot

These species tolerate very poor water–they have adaptations to obtain oxygen.

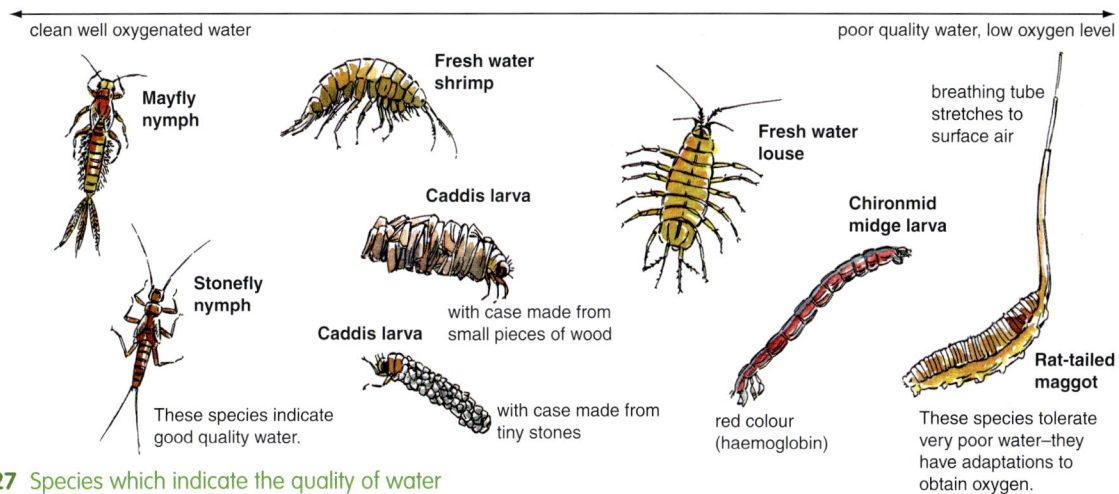

Figure 3.27 Species which indicate the quality of water

Summary

✓ Animals must obtain all they need to eat, grow and reproduce from the environment in which they live.

✓ Animals have features called **adaptations** which allow them to survive in the environment in which they live. These features enable them to survive at the temperature of the area, find food within the area and locate a breeding site.

✓ **Population size** is controlled by **predation** and **competition** for food, water and breeding sites.

✓ Plants growing in dry areas have adaptations to reduce water loss, such as thick waxy cuticles, water storage tissue and long roots.

✓ Plant populations are controlled by the numbers of herbivores (grazing) and by competition for light, water and minerals.

✓ **A balanced ecosystem** changes slowly and the numbers of producers, herbivores and carnivores remain fairly constant.

✓ **Biological control organisms** such as ladybirds, insectivorous birds and ground beetles control pest species, such as aphids. Kestrels and owls control mice.

✓ The increasing human population has an impact on the environment through the use of more land for buildings, roads, intensive farming and waste disposal.

✓ The increasing human population uses more water and raw materials.

✓ The increasing human population produces more air pollution, more water pollution, and more waste.

✓ **Deforestation** for timber production, arable farming and cattle ranching can have damaging effects on the whole environment including the land, the wildlife and the human population.

✓ **Global warming** is the result of increased carbon dioxide and methane levels in the atmosphere.

✓ Conclusions linking global warming and greenhouse gas levels require careful measurements over many years, not anecdotal evidence.

✓ In environmental situations where many variables are possible, scientists construct computer models to manipulate the data and make predictions for the future. This applies to predictions about global warming.

✓ The consequences of global warming are climate changes which result in the flooding of low-lying areas and change in the direction of rain-bearing winds.

✓ Action taken by the public to reduce waste, recycle more, and use energy efficiently will benefit society and the environment.

✓ **Sustainable development** is necessary to make sure that fuel, food and a pleasant environment are available for future generations.

✓ Environmental monitoring can be carried out through chemical sampling, but **bioindicators** of water or air pollution give a longer term view of changes.

✓ Sampling of bioindicators requires a well-planned investigation and data collected over many years for comparison.

1 An animal living in the Arctic is likely to have:
A thick fur, thick fat layer, long thin legs, white fur;
B thick fur, thick fat layer, short thick legs, white fur;
C thick fur, thick fat layer, long thin legs, dark fur;
D thick fur, thin fat layer, short thick legs, white fur. *(1 mark)*

2 The factors which would maintain low population numbers of robins are:
A cold winters, shortage of nest sites, local cats which hunt, warm spring with many insects;
B mild winters, shortage of nest sites, local cats which hunt, warm spring with many insects;
C mild winters, shortage of nest sites, local cats which hunt, cold spring with few insects;
D cold winters, shortage of nest sites, local cats which hunt, cold spring with few insects. *(1 mark)*

3 Which statement best describes biological indicator species?
A Living organisms which can survive in a wide range of conditions.
B Fossil remains that are found in sedimentary rocks.
C Living organisms that are sensitive to changes in the environment.
D Living organisms that change colour to suit their background. *(1 mark)*

4 Which definition best describes sustainable development?
A Development which provides for present needs but may change the environment.
B Development which provides for future needs by changing the environment.
C Development which has low environmental impact and provides for present and future needs.
D Development which has high environmental impact and provides for present and future needs. *(1 mark)*

5 As part of the policy for sustainable development the government proposes 'a reduction in development on "green field sites" and increased use of "brown field sites"'.
a) What is meant by the term 'sustainable development'? *(2 marks)*
b) How will this proposal benefit the environment? *(3 marks)*

6 Four groups of students investigated the invertebrates in West Brook, a shallow stream. Kick-sampling was used to obtain samples. Groups A and B were sampling on the upstream side of a row of cottages, where the water was clear and the stones on the bottom could be seen easily (Figure 3.28). Groups C and D sampled on the downstream side from the cottages. Here the water was less clear and the stones were covered with a black slimy deposit.

Before starting the groups discussed how they

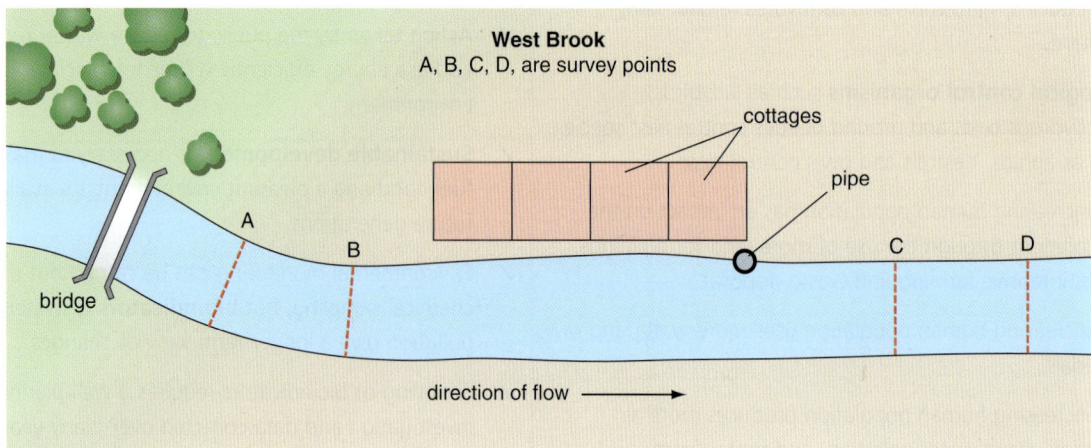

Figure 3.28 A map of the survey sites and cottages at West Brook

would make their investigations a 'fair test'. The results of their samples are shown in Table 3.3.

a) Copy and complete the bottom row in the table and calculate the average number of species in the upstream samples and downstream samples from the cottages.

(3 marks)

b) What should the four groups do to make 'fair tests' and use their sampling results to compare the two parts of the stream?

(1 mark)

c) Describe the evidence in the table which indicates that the upstream region was 'healthy'. *(3 marks)*

d) What could have happened by the cottages to change the life in the stream? Give two pieces of evidence which support your suggestion. *(3 marks)*

Organisms	Oxygenation of water	Group A	Group B	Group C	Group D
Stonefly nymphs	Only survive in well oxygenated water	+++++	+++++		
Mayfly nymphs	Only survive in well oxygenated water	+++	+++++		
Dragonfly nymphs	Survive in oxygenated water		+		
Small fish	Survive in oxygenated water	+			
Freshwater shrimps	Require a reasonable level of oxygenation	+++++	+++++		+
Freshwater lice	Can survive in low levels of oxygenation	+++	+++	+++	+++
Caddis with vegetation case		+	+		
Caddis with stone case			+		
Daphnia		+++++	+++++		
Water snails		+++++	+++++		
Water boatman beetles		+++	+		
Small worms		+++++	+++++		
Midge larvae (red)	Survive at low oxygen levels	+	+	+++++	+++++
Rat tailed maggots	Survive at low oxygen levels			+++++	+++++
Total number of species					

Table 3.3 The results of an invertebrate survey of West Brook
Key: + indicates one specimen, +++ several, +++++ large numbers.

Chapter 4
How can we explain reproduction and evolution?

At the end of this chapter you should:

✓ be able to explain how genes are passed from parents to their offspring;

✓ know where the genes that carry genetic information are found;

✓ be able to describe the differences between sexual and asexual reproduction;

✓ understand the different techniques used to produce clones of animals and plants;

✓ be able to judge the economic, social and ethical issues concerning cloning and genetic modification;

✓ be able to evaluate different theories of evolution;

✓ be able to explain why Darwin's theory of evolution is now the most widely accepted;

✓ know how fossils provide evidence for the evolution of different organisms;

✓ be able to explain how natural selection has led to the evolution of new species;

✓ be able to suggest why different species became extinct.

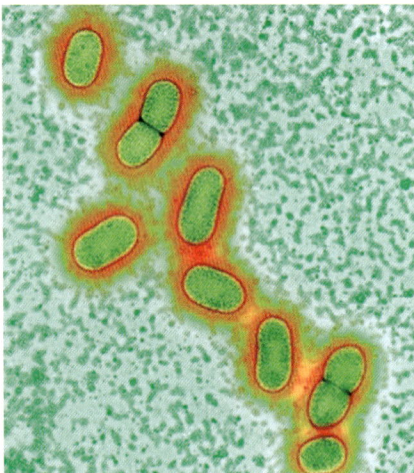

Figure 4.1 In all these photographs characteristics have been inherited from one generation to the next

4.1 How did an Austrian monk help to shape our understanding of inheritance?

We now know how the features and characteristics that we inherited from our parents are passed on, but this was not always the case. It was only through careful experimentation that scientists gained this understanding. In this section, you will learn about the key developments that led to our current understanding of inheritance.

For many years people believed that the characteristics of parents combined in some way when they had children. For example, if a mother had black hair and a father had blonde hair, the two colours would combine to produce brown-haired children. This was also thought to be true of animals and plants. It was known as the 'blending theory' of inheritance.

An Austrian monk called Gregor Mendel, who was born in 1822, disagreed with this theory. Mendel worked in the gardens of an Austrian monastery and noticed that pea plants had either purple or white flowers. When they produced seeds the new plants that grew only ever had purple or white flowers, never a mixture of the two colours. He concluded that the parent plants passed on specific features to the offspring plants and not combined features.

Although Mendel had not trained as a scientist, he realised that he needed to carry out controlled experiments. He could then collect reliable evidence to prove his theory. He started by doing some

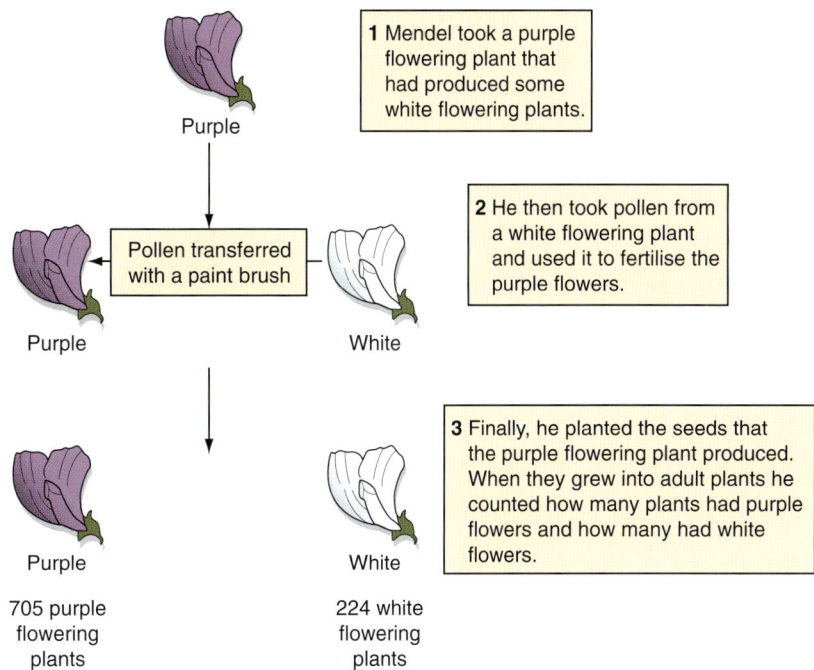

Figure 4.2 Gregor Mendel at work

1 Mendel took a purple flowering plant that had produced some white flowering plants.

Purple

Pollen transferred with a paint brush

Purple

White

2 He then took pollen from a white flowering plant and used it to fertilise the purple flowers.

Purple

White

705 purple flowering plants

224 white flowering plants

3 Finally, he planted the seeds that the purple flowering plant produced. When they grew into adult plants he counted how many plants had purple flowers and how many had white flowers.

Figure 4.3 A flow diagram showing the procedure Mendel used to investigate inheritance in pea plants and the results he collected

❶ What does the word 'inheritance' mean when it is used in science?

❷ What prompted Gregor Mendel to carry out his research into inheritance?

❸ What did Mendel do to ensure that his findings were reliable?

❹ Why was it important for Mendel to obtain reliable results?

❺ Explain how Mendel's results showed that the blending theory of inheritance was unsatisfactory.

❻ a) Why do you think it took so long for Mendel's ideas to be accepted?
   b) Explain why the work of other people helped Mendel's explanation to replace the blending theory.

preliminary experiments in which he pollinated purple and white flowered plants together. From this he confirmed that when these two types of pea plants were crossed they produced plants with white flowers and plants with purple flowers but never any plants with colours in between.

For his main experiment, Mendel took a purple flowering plant, that had produced some white flowered plants, and a white flowering plant. He then pollinated these two plants together for several generations and counted the numbers of purple and white flowering plants that were produced.

Mendel analysed his results and discovered that for every three purple flowering plants there was only one white flowering plant, a ratio of 3:1. From his results, Mendel concluded that some inherited characteristics, such as purple flowers, had a stronger influence over the offspring. He called these 'stronger' influences **dominant** and the 'weaker' influences **recessive**. These findings showed clearly that the blending theory was insufficient to explain inheritance.

Mendel published his findings in 1866 expecting people to appreciate that **inheritance** could be explained much better using his ideas. Sadly, his ideas were largely ignored for 34 years until three other researchers published similar findings based on their own experiments. At the same time, Mendel's work was translated into English and people started to take his ideas seriously. They realised that Mendel provided a far better explanation of inheritance than the blending theory.

4.2 Cell structure and inheritance

Mendel pointed out that different characteristics were passed from parents to their offspring. He also showed that some characteristics were dominant and some recessive, but he couldn't explain *how* these characteristics were passed on. It was not until research on cell structure was carried out that the process of inheritance could be explained.

The information that controls inherited characteristics is carried by **genes**. Genes are sections of **chromosomes**, which are long polymer molecules made of deoxyribonucleic acid, or DNA. Chromosomes are found in the nuclei of all cells. There are 46 chromosomes in a normal human cell. These 46 chromosomes carry about 25 000 genes.

Different genes control different characteristics. For example, the flower colour that Mendel studied was controlled by one gene in the pea plant, while the seed colour was controlled by a different gene. Eye colour in humans is also controlled by one gene. These genes are passed on from parents to their children and are carried in the nuclei of sex cells during reproduction. The male sex cells in animals are the sperm cells and in

Inheritance is the passing on of characteristics and appearance from parents to their offspring.

Chromosomes are polymers of DNA. They are found in the nuclei of all cells.

A **gene** is a section of a chromosome that carries information for a certain characteristic.

Gametes are the sex cells: sperm, egg cells and pollen.

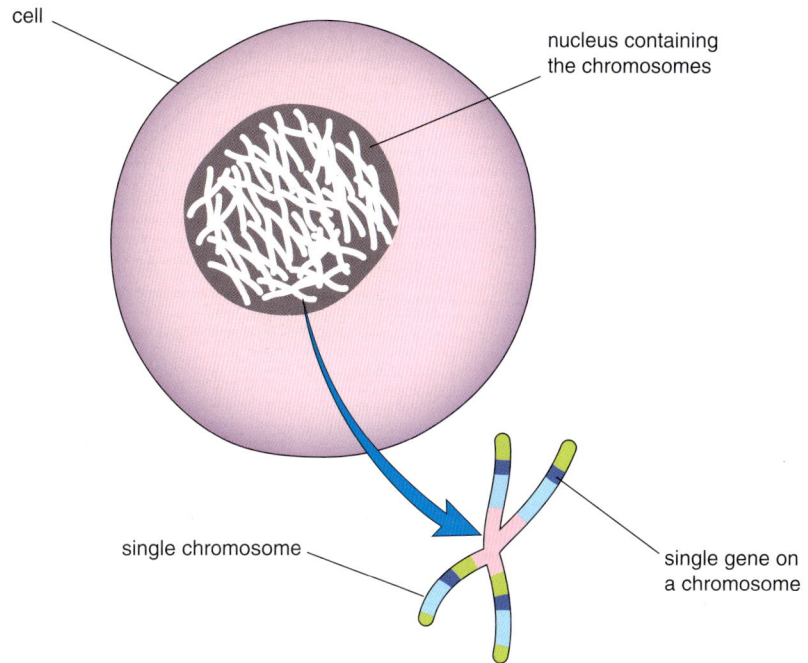

cell

nucleus containing the chromosomes

single chromosome

single gene on a chromosome

Figure 4.4 A diagram showing where genes are found within a cell

An egg cell from the mother.

The *nucleus* in each cell contains half the genes needed to form a baby.

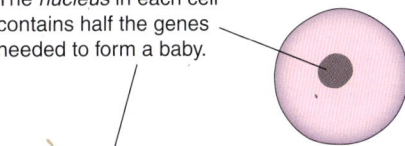

A sperm cell from the father.

The sperm cell enters the egg cell when *fertilisation* takes place.

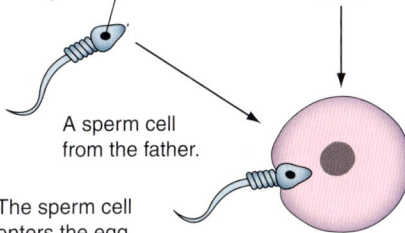

The two nuclei start to fuse together.

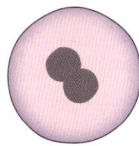

The cell that is formed is called a *zygote*. The nucleus now contains all the genes needed to form a baby.

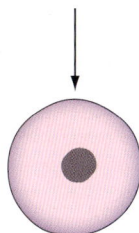

Figure 4.5 A flow diagram showing how genes are inherited by offspring from their parents

plants they are pollen cells. Female sex cells are called egg cells. A general name for sex cells is **gametes**.

Sexual reproduction takes place in animals and plants and involves two parents. During sexual reproduction, a male sex cell fuses (joins) with a female sex cell, to form one new cell. The new cell gets half its genes from one parent and half its genes from the other parent. This new cell develops into a new plant or animal. The offspring's development is controlled by the genes that it has inherited from its parents. Since the offspring has inherited half its genes from each parent, both parents influence its appearance and characteristics.

Have you noticed that children in the same family are often very different in looks and have totally different abilities even though they have the same parents? Scientists call these differences **variation**. Variation occurs because different genes from each parent can combine when fertilisation takes place. This gives each child in a family a different set of genes unless they are identical twins. It's a bit like selecting five cards from each of two packs. You would almost always end up with two different sets of five cards. As there are about 25 000 genes in a human cell, the combinations are almost endless.

Sexual reproduction is the joining of male and female gametes. It results in variation in the offspring of parents.

Asexual reproduction produces genetically identical offspring from only one parent. It does not inolve gametes.

7 What are 'inherited characteristics'?

8 a) Where are genes found?
   b) What do genes do?

9 Why do you think humans need a large number of genes? Explain your answer.

10 How are characteristics inherited by children from their parents?

11 Why does sexual reproduction result in variation of children in the same family?

4.3

How has an understanding of asexual reproduction helped with cloning?

Asexual reproduction does not involve gametes and only one parent is needed. This means that all the genes in the offspring come from just one parent. In fact, the genes in the offspring are exact copies of those in its parent. So, **asexual reproduction** produces offspring that are genetically identical to their parents. These genetically identical offspring are known as **clones**.

Cloning is not as new as you may think. Although there is a lot of media 'hype and interest in cloning, plant growers have been using cloning techniques for a long time. Some plants and animals reproduce naturally through asexual reproduction, producing clones of themselves. For example, strawberry plants produce runners, which are stems that grow along the surface of the ground. These runners produce roots that grow down into the soil and form a new plant. As the new plant has been produced entirely from one parent, it has exactly the same genes as the parent plant. Bacteria, which are single celled organisms, also reproduce asexually. They copy all their genes and then split in half with a complete set of copied genes going into the new cell. Again, this produces a new individual with identical genes to its parent.

The process of asexual reproduction is used by plant growers to produce cheap, genetically identical new plants by taking **cuttings** from older plants. *Peperomia* are often bought as house plants. New *Peperomia* can easily be produced from older plants by taking cuttings. This is often carried out in a commercial nursery to mass produce new plants.

Figure 4.6 Asexual reproduction of a strawberry plant and a bacterium. In each case the offspring is genetically identical to its parent.

Cloning refers to techniques that are used to produce genetically identical individuals.

Cuttings are taken from plants to produce new, genetically-identical plants.

Figure 4.7 A *Peperomia* house plant produced from a cutting

⑫ What are the key differences between sexual and asexual reproduction?

⑬ What is a clone?

⑭ a) What are the benefits to horticultural companies of producing new plants by taking cuttings from older plants?
b) What problems do you think may result from taking cuttings?

⑮ Roots often grow from the cut stem of a plant after a cutting has been taken.
a) What does this suggest about the cells in the stem of the plant?
b) What do you think is the purpose of the hormone rooting powder that cut stems are dipped in?

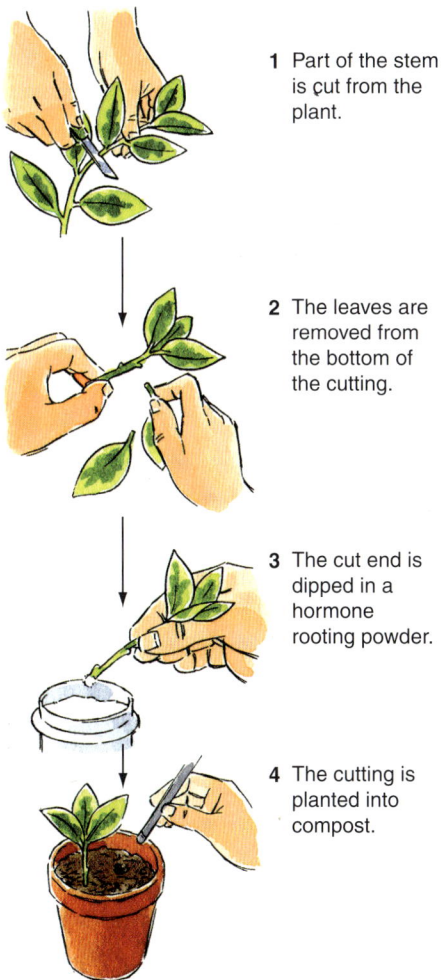

1 Part of the stem is cut from the plant.

2 The leaves are removed from the bottom of the cutting.

3 The cut end is dipped in a hormone rooting powder.

4 The cutting is planted into compost.

Figure 4.8 A flow diagram showing how a *Peperomia* plant is grown from a cutting

Activity – Modern cloning techniques

Cloning is regularly in the news with headlines such as:

'Cloned cows may be safe to eat.'

'Korean and US scientists claim human cloning breakthrough.'

'UK court ruling means cloning not illegal.'

'After Dolly the sheep comes Snuppy the puppy.'

It is important that you understand the science behind headlines like these in order to form your own views about the media coverage of topics like cloning.

In this activity you will be looking at information about three modern cloning techniques and then considering their potential uses. You will also need to think about the issues that these developments present to society.

Tissue culture or micro-propagation is a way of producing large numbers of plants very quickly. A small number of cells are taken from a 'parent' plant and grown in a medium which is rich in nutrients and plant growth hormones (Figure 4.9).

Embryo transplant is a way of splitting the embryo from a pregnant animal and then transplanting the divided clumps of cells into a number of host mothers (Figure 4.10).

Fusion cell and adult cloning is used to produce exact copies of cells or whole individuals from a single cell (Figure 4.11).

❶ Each flow diagram below shows a different cloning technique. For each technique describe how genetic information is passed from the parent plant or animal to the clone.

❷ Look carefully at Table 4.1. This shows three benefits and three drawbacks of using tissue culture to clone plants.

❸ a) In pairs, discuss the benefits and drawbacks of embryo transplant techniques.
  b) Draw a table showing three benefits and three drawbacks of embryo transplant techniques.

❹ a) In pairs, discuss the benefits and drawbacks of using fusion cell and adult cloning techniques.

b) Draw a table showing three benefits and three drawbacks of fusion cell and adult cloning techniques.

❺ The pictures in each of the flow diagrams show a current application of each cloning technique. Describe briefly another possible application for:
  a) tissue culture;
  b) embryo transplant;
  c) fusion cell and adult cloning.
  You may need to do some research to complete this question. A search on 'cloning' at www.bbc.co.uk will give you lots of examples.

❻ '*Research into cloning humans is wrong and should be banned.*'
  This statement is a view held by many people. You are going to prepare a mini-debate about this viewpoint.

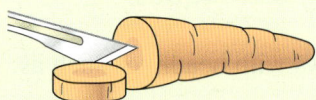

1. A sample of plant tissue is removed from the parent plant.

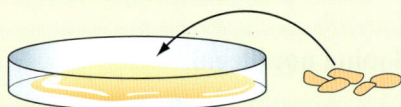

2. The tissue sample is cut into small pieces and placed in a dish of agar jelly containing nutrients and plant growth hormones.

3. Each piece of the tissue sample grows into a clump of cells which develop into small, individual plants called plantlets.

4. The plantlets are transferred into soil or compost where they grow into adult plants.

Figure 4.9 Tissue culture is used to produce large numbers of plants from one parent plant

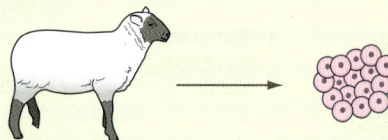

1. An embryo (clump of developing cells) is removed from a pregnant animal.

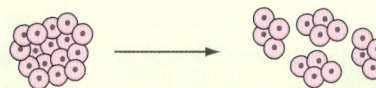

2. The embryo is split into a number of smaller clumps of cells.

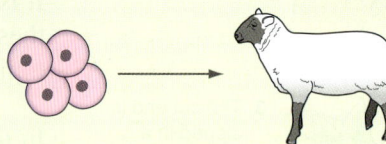

3. Each new embryo is inserted into the uterus of another host mother.

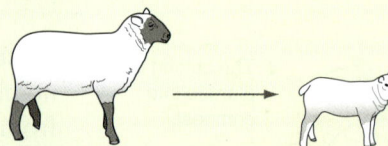

4. Some of the host mothers become pregnant and give birth to their offspring.

Figure 4.10 Embryo transplant is used to implant divided clumps of cells (new embryos) from one embryo into a number of host mothers

Benefits of tissue culture	Drawbacks of tissue culture
A lot of new plants can be grown in a relatively short time	All plants have the same genes, so they will all be vulnerable to the same diseases or pests
Little space is needed, and conditions can be precisely controlled	There is no way that new beneficial characteristics can arise by chance
All new plants inherit the same characteristics	The absence of variation in the plants increases the danger of reducing the gene pool

Table 4.1 Three benefits and three drawbacks of using tissue culture to clone plants

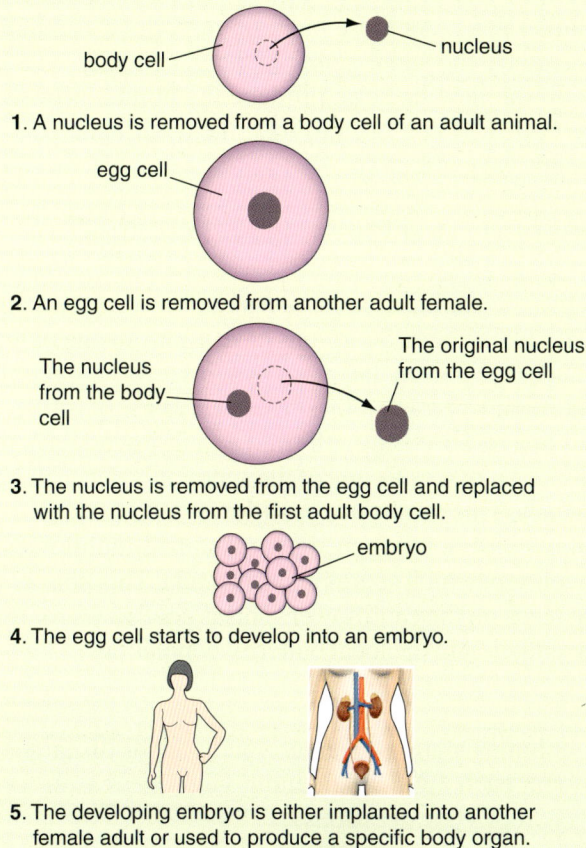

body cell — nucleus

1. A nucleus is removed from a body cell of an adult animal.

egg cell

2. An egg cell is removed from another adult female.

The nucleus from the body cell

The original nucleus from the egg cell

3. The nucleus is removed from the egg cell and replaced with the nucleus from the first adult body cell.

embryo

4. The egg cell starts to develop into an embryo.

5. The developing embryo is either implanted into another female adult or used to produce a specific body organ.

- Organise yourself into a group of five. Two people will argue *for* the statement, two will argue *against* the statement and the fifth person will act as a judge.
- Each of the pairs should research and prepare a speech supporting their argument. The judge should research the topic fully and become an expert on cloning.
- Each pair will then deliver their speech to the other pair whilst the judge notes down the main points that are raised.
- The judge will then decide which pair have made the most convincing argument.
- Finally, each judge gives feedback to the whole class stating the main points raised by each pair and giving his / her verdict.

Figure 4.11 Fusion cell and adult cloning is used to produce cloned animals and human body organs

4.4 Understanding genetic engineering

Research involving genetic engineering has been going on for over 40 years. Recent developments show that the techniques could have many useful applications. However, many people are seriously concerned about the possible risks and ethical issues linked to genetic engineering.

Genetically engineering the human growth hormone

Children whose pituitary gland does not produce enough growth hormone suffer from a condition called pituitary dwarfism. They grow very slowly and often reach puberty long after others of the same age. Until the mid-1980s the condition was treated with growth hormone

Figure 4.12 A diabetic injecting herself with genetically engineered insulin. In the past, insulin had to be taken from slaughtered animals

Figure 4.13 A protest against genetically modified crops

1. The gene that produces growth hormone is removed from the nucleus of the cell.

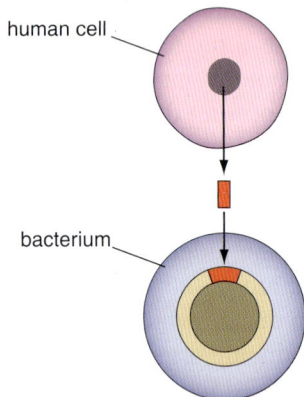

human cell

bacterium

2. The gene is then inserted into the DNA of a bacterium. The bacterium then starts producing human growth hormone.

Figure 4.14 A flow diagram showing how genetically engineered human growth hormone is produced

taken from the corpses of unaffected people. This procedure came with a serious risk. Some children treated with the growth hormone developed a disease of the nervous system called CJD. The use of human growth hormone was stopped when this was discovered.

Human growth hormone is now produced by genetic engineering. Chromosomes are extracted from a normal human body cell. The gene that produces the growth hormone is 'cut out' from the chromosomes using an enzyme. The human growth hormone gene is then inserted into the DNA of a bacterial cell. The bacterium then multiplies making lots of bacteria with the human gene. As these bacteria reproduce asexually, each new bacteria has exactly the same genes as the genetically modified bacteria. So, all the new bacteria produce the human growth hormone which can be collected and used to treat the dwarfism. The treatment is expensive but very effective.

Genetically engineering cauliflowers to kill caterpillars

Another application of genetic engineering involves cauliflowers that can kill the caterpillars that try to eat them. Scorpions have a gene in one of their chromosomes that produces the poison they use to kill insects. This gene can be removed from the scorpion's chromosomes with an enzyme. It is then inserted into the nucleus of a fertilised egg cell from a cauliflower. Each time the fertilised egg cell divides it makes a copy of the scorpion's gene, along with copies of its own genes. This results in a cauliflower with the poison producing gene in all of its cells. So, the genetically modified cauliflower now has a poison in its cell sap, which acts as an insecticide. This poison kills any caterpillar that feeds on the cauliflower. This is good news for the farmer, but it leads to a reduction in the local butterfly population.

16 Explain the term 'genetic engineering'.

17 Some parents of children suffering from pituitary dwarfism refused the treatment with growth hormone from human corpses. Suggest two reasons why they refused this treatment.

18 Suggest two benefits of using genetically engineered human growth hormone instead of growth hormone from human corpses.

19 The poison from genetically modified cauliflowers does not harm humans when it is eaten. Why do you think people still refuse to eat the genetically modified cauliflower?

20 Table 4.2 lists a number of applications of genetic engineering. Copy and complete the table adding columns stating whether you agree or disagree with each application and giving one reason for each of your decisions.

Application of genetic engineering
Producing human insulin from genetically modified bacteria
Producing disease resistant rice to increase its yield in developing countries
Producing chickens with four legs and no wings to increase meat production
Parents being able to choose the sex of their baby
Developing plants that can produce plastics to reduce the use of crude oil

Table 4.2

4.5 How do new species of plants and animals evolve?

We live on a planet with a fantastic variety of different animals and plants. Scientists have estimated that there are between 2 million and 100 million different species of plants and animals on the Earth today, but they don't know the exact number. How did all these different animals and plants come to be living on Earth? Have there always been similar plants and animals on Earth? Were the present plants and animals like their predecessors (forerunners) or very different from them? Although there are answers to some of these interesting questions, scientists cannot be certain how life began on Earth.

Most scientists believe that the plants and animals alive today have developed from forerunners that lived in the past. These forerunners were not the same as today's species. Changes have taken place over many generations. Scientists say that the present plants and animals have **evolved** from their forerunners and they call the overall process **evolution.** This evolution has resulted in species that are adapted to survive in the environments in which they live. No one has actually seen these changes taking place because the process takes so long. But there have been plenty of theories that try to explain evolution.

Figure 4.15 Some of the species of animals and plants found on the Earth today

Activity – Theories for evolution

In this activity you will be considering five different theories for the evolution of plants and animals. Think about the way in which the theories conflict with each other and try to decide which one offers the most convincing explanation.

Creationism is based on the idea that every species was created separately by God and that species do not change through time. This was the view held by the Christian Church in the seventeenth century and strongly reinforced by Archbishop James Ussher. Ussher studied the Bible and calculated that all life had been created in 4004BC. According to his theory, all species alive today were created about 6000 years ago in the same form as they are now. As plants and animals reproduced, there were no changes from generation to generation.

A second theory of evolution was proposed by the French zoologist, George Leclerc **Buffon**. In the late eighteenth century, a number of European scientists

Figure 4.16 Archbishop James Ussher (1581–1656) believed strongly in creationism

began to question the creationism idea that species had never changed. Buffon actually suggested that the Earth was at least 75 000 years old and that certain species had changed in that time. He explained that these changes were sometimes influenced by the environment and sometimes happened by chance. He even suggested that humans and apes were related. Buffon's anti-creation ideas were kept quiet, and simply recorded in a book called *Histoire Naturelle* which sold in only limited numbers.

Figure 4.17 George Leclerc Buffon (1707–1788) was the first scientist to suggest that species might evolve over thousands of years

In the early nineteenth century, a biologist called Jean Baptiste **Lamarck** published his theory of evolution. He suggested that microscopic organisms could form from non-living material and then gradually evolve into more complex species, eventually resulting in humans. Lamarck believed that these changes happened during an animal's lifetime. The changes that an animal developed were then passed on to its offspring. For example, he explained that herons had evolved long legs because individual herons had stretched their legs to stay dry when wading in water. These herons would then pass on this characteristic to their offspring who would, in turn, stretch their legs even more and produce offspring with even longer legs. He also used this theory to explain why giraffes have such long necks.

Lamarck's theory was severely criticised by another French scientist called George **Cuvier.** Cuvier argued that changes to an individual animal during its life could not be passed on to its offspring. For example, a person who trains hard will develop bigger muscles but will not then have children that grow up having bigger muscles. Cuvier did however agree with Lamarck that there had been different species in the past. Cuvier's theory was called **catastrophism.** It suggested that sudden environmental changes, such as earthquakes, killed whole species. After catastrophes such as this, new species slowly took their places. Cuvier based his ideas on fossil records.

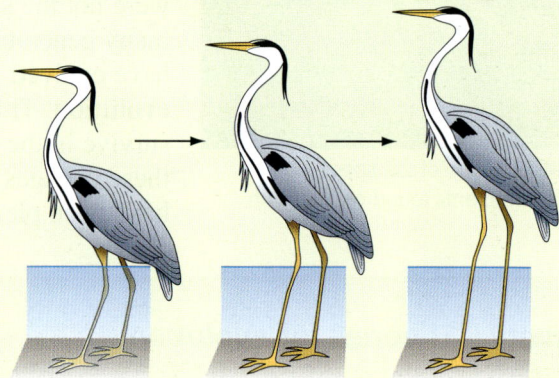

Figure 4.18 Lamarck's explanation for the heron's long legs

The final theory for you to consider is that proposed by the Englishman, Charles **Darwin**, in the nineteenth century. Darwin studied theology but was more interested in biology. He joined the crew of a ship called the Beagle on a five-year voyage to Africa, South America and Australia. During this trip, Darwin studied the wildlife on the Galapagos Islands, which are 1000 kilometres from the west coast of South America. He found that the islands had species of animals that were not found anywhere else in the world. For example, he found 13 species of finches on the islands whereas only one of these species lived on the mainland of South America. The different species of finch on the Galapagos Islands had differently shaped beaks that suited the food they ate. For example, one species had a very strong beak for cracking nuts while another, living on a different island, had a long thin beak for picking up insects.

Darwin suggested that the different finches on the islands had evolved from a single species that had migrated from the mainland. He explained that within a population of finches there were always some variations in features, such as beak size and shape. Individual birds with features that better suited their environment were more likely to survive and pass on these features to their offspring. Darwin found a similar pattern to the finches in other species on the Galapagos islands.

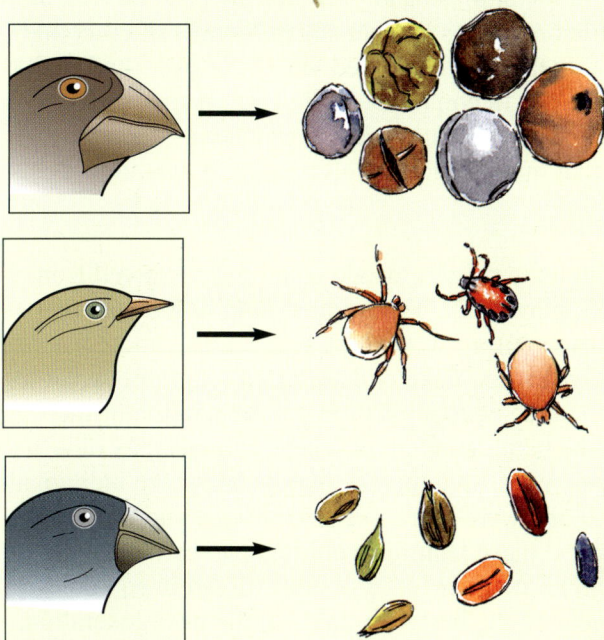

Figure 4.19 Three finches found on the Galapagos Islands and the food they eat

❶ Copy and complete Table 4.3 summarising the main points in each theory. The first one has been done for you.

Theory or scientist	Main points
Creationism	• God created all species alive today about 6000 years ago • Species have not changed over time
Buffon	
Lamarck	
Cuvier / catastrophism	
Darwin	

Table 4.3

❷ Why do you think Buffon chose not to publish his ideas?

❸ Explain the problems with Lamarck's theory, using your understanding of genes and inheritance. If you need help, read through Section 4.2.

❹ Darwin's theory of evolution is now widely accepted.
  a) What are the main differences between Darwin's ideas and the other theories?
  b) Why do you think Darwin's ideas give a better explanation of evolution?

Figure 4.20 This drawing of Darwin's head on a monkey's body was printed after he published his theory of evolution

❺ Darwin's ideas were only gradually accepted. Look at Figure 4.20. Why do you think a cartoon like that in Figure 4.20 was printed?

❻ Creationism reflected a religious belief about life on Earth. Today, Christians still believe that God created life and many of them also accept Darwin's theory of evolution. Discuss in pairs how Christians can believe that God created life on Earth and also accept Darwin's theory of evolution.

Evidence for evolution and natural selection

Darwin formed his theory of evolution from his observations on the Galapagos Islands and elsewhere. Today we have much more evidence to support his theory. The theory of evolution states that all species have evolved from simple life forms which first appeared on the Earth 3000 million (3 billion) years ago. So how do we know what the forerunners of today's species were like? How did we get our evidence for evolution?

Most of our evidence for evolution comes from fossils. Scientists study fossils to find out about the animals and plants of the past. Fossils show how much or how little organisms have changed. Fossils of crocodile jaws and teeth show that they have evolved very little over the last 200 million years (Figure 4.21). In comparison, the earliest fossils of our human ancestors, such as *Orrorin tugenensis*, which was the size of a chimpanzee, show that humans have evolved a great deal in just 2 million years.

Fossil records show how the modern whale could have evolved from a land-based mammal (Figure 4.22). The original land-based mammal from which whales have evolved was called Pakicetidae, about the size of a wolf. Millions of years ago Pakicetidae started to go into water to find a more abundant source of food. Within any population of the animals, some would be better swimmers. These animals may have had wider feet or have been more streamlined. As a result, these individuals could catch more food and were more likely to survive and reproduce. The genes that controlled these features would be passed on to their offspring who would inherit wider feet and more streamlined bodies. These individuals would be the best swimmers in the next generation. Over millions of generations of gradual change, it is possible that Pakicetidae was transformed into the modern whale. Of course, this is only a theory but it is supported by fossil records.

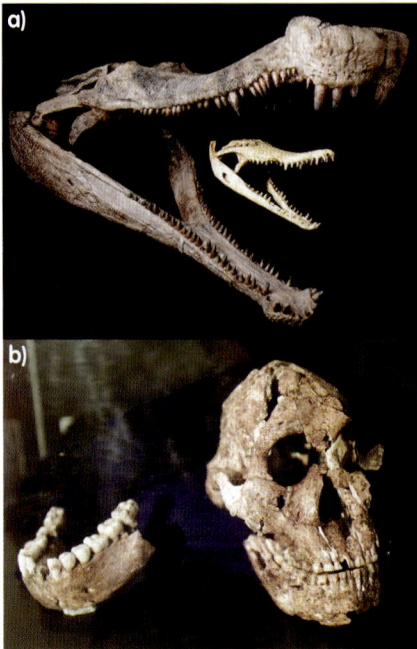

Figure 4.21 Fossils of a) a crocodile's jaw and teeth from 200 million years ago and b) the bones of *Orrorin tugenensis* from 2 million years ago

㉑ What information about life on Earth can be obtained by studying fossils?

㉒ Why is it important to have a series of related fossils when studying evolution?

㉓ Fossil records for the evolution of whales provide evidence to support the theory of evolution, but they do not prove it. Explain why this is so.

24 It has been suggested that today's horses evolved from an ancestor the size of a fox. Write the following sentences in the correct order to explain how the horse may have evolved by natural selection.

- Gradually, over many generations, taller horses with the toes joined together were produced.
- Some *Eohippus* had slightly longer legs than others and could run faster.
- Some *Eohippus* also had toes that were closer together, which helped them to run even faster.
- Eventually the modern horse evolved with long legs and hooves instead of toes.
- The individuals that survived were more likely to reproduce and pass on their characteristics to their offspring.
- Sixty million years ago, a small mammal, *Eohippus* existed that had three toes and short legs.
- These individuals were better at avoiding predators and were more likely to survive.

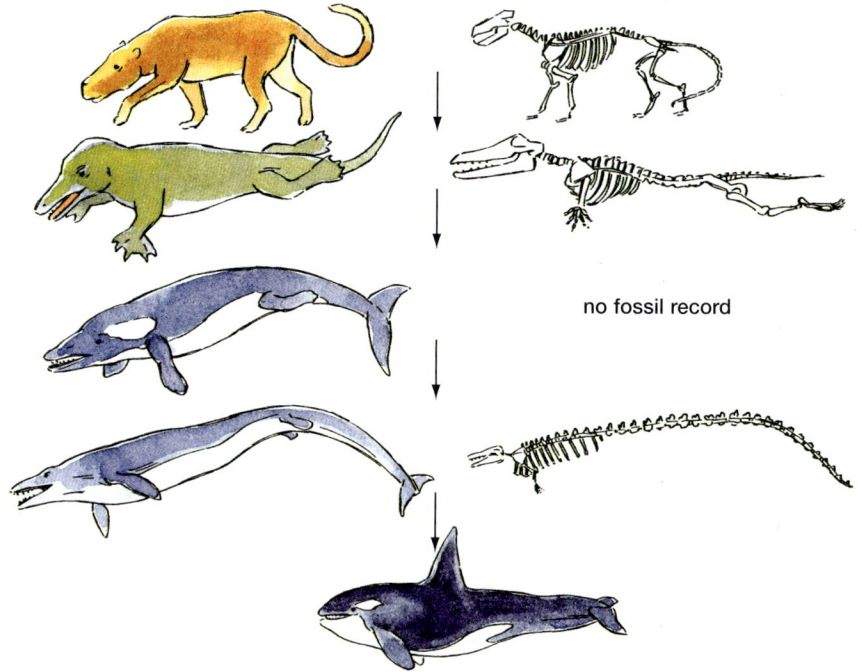

no fossil record

Figure 4.22 Pictures showing how the whale may have evolved from a land-living mammal

The sequence of events that leads to evolution of a species is called **natural selection**. Natural selection can be split into five key steps.

1 The individuals within a species vary because of differences in their genes. These variations result from the inheritance of different genes or from the occasional mutation (change) of genes.
2 Individuals with the characteristics that best suit their environment are more likely to survive.
3 These individuals are also more likely to reproduce successfully.
4 The genes which have enabled these individuals to survive are then passed on to the next generation.
5 This process is repeated over many generations leading to a new species.

4.7 Why do species become extinct?

An animal becomes extinct when there are no more living members of its species. Extinction is often part of evolution. As a new species evolves, the old one dies out. Sometimes a species becomes extinct without a new species evolving. The main causes of extinction are:
- changes to the environment;
- new predators;
- new diseases;
- new competitors.

It is difficult to know the cause of an extinction when it occurred so long ago. The dodo (Figure 4.23) was a flightless bird that lived on the

Figure 4.23 The dodo became extinct around 1700.

㉕ Why are we still not sure what caused the dodo to become extinct?

㉖ In the last one hundred years, humans have been responsible for a large increase in the number of species that have become extinct.
  a) Which human activities have caused this increase in extinction rates?
  b) What problems could this loss of species cause in the future?

island of Mauritius. It finally became extinct around 1700 after humans had settled on the island. For many years, scientists thought that the dodo became extinct because the people that moved to the island hunted and ate the dodo. In other words, humans were a new predator to the dodo. More recently, however, records suggest that the dodo was not widely eaten by the settlers.

When humans moved to Mauritius, they cut down most of the forests to make space for their homes. In doing so, they destroyed the habitat in which the dodos lived. As the dodos were flightless, they could not easily migrate to a different island. So, changes in the environment may have resulted in extinction of the dodo.

The settlers brought a number of animals with them that were new to Mauritius. These included monkeys, pigs and rats. These animals ate the same food as the dodos and destroyed their nesting sites. The new animals competed with the dodos for habitat and food, which might also explain why the dodos became extinct.

Summary

✓ Some characteristics are passed on from parents to their offspring. This is called **inheritance**.

✓ The information that controls inherited characteristics is carried by **genes** which are passed on in the sex cells (**gametes**).

✓ A gene is a section of a **chromosome**. Chromosomes are found in the nuclei of all cells.

✓ **Sexual reproduction** is the joining of a male and a female gamete. Sexual reproduction involves the mixing of genetic information from two parents and leads to variety in their offspring.

✓ **Asexual reproduction** only requires one parent and does not involve the joining of gametes. Offspring produced by asexual reproduction from one parent are genetically identical and show no variation.

✓ Genetically identical individuals are called **clones**.

✓ New plants can be produced quickly and cheaply by taking cuttings. This is a type of cloning.

✓ Modern cloning techniques include tissue culture, embryo transplants and fusion cell and adult cloning.

✓ It is important to be able to make informed judgements about the economic, social and ethical issues concerning the use of cloning and genetic engineering.

✓ Genetic engineering can be used to transfer genes from the cells of one organism to the cells of other organisms.

✓ Genes can be transferred to the cells of plants or animals to introduce new characteristics to the organism.

✓ There have been a number of conflicting theories to explain evolution.

✓ Darwin's theory of evolution is now the most widely accepted. Darwin's theory explains how the evolution of new species occurs by natural selection.

✓ The theory of evolution states that all species have evolved from simple life forms which first appeared more than 3 billion years ago.

✓ Fossils provide us with strong evidence for the theory of evolution.

✓ The extinction of a species may be caused by: changes to the environment, new predators, new diseases or new competitors.

✓ It is not always easy to pinpoint what caused a particular species to become extinct.

1 Young rabbits, like these, often look like their parents. This is because information about their appearance, for example fur tone, is handed down from parents to their offspring.

Copy and complete the paragraph below using the appropriate words from the box.

| body | chromosomes | clones |
| cytoplasm | genes | nucleus | sex |

Information is passed from parents to their offspring in _____ cells. Appearance is controlled by _____. The structures which contain information about a large number of characteristics are known as the _____. In the cell these structures are found in the _____. *(4 marks)*

2 Humans reproduce by sexual reproduction. Michael and James are brothers, but they have very different features.

a) State two ways in which sexual reproduction is different from asexual reproduction. *(2 marks)*

b) Explain why Michael and James have different features. *(3 marks)*

3 Read the piece of text below carefully.

A woman from Texas ordered a genetic replica of her cat, Nicky, when he died aged 17. The new cloned cat, named Little Nicky, who cost his new owner $50 000, was created using DNA taken from his namesake, Nicky. 'He is identical. His personality is the same' said the owner. During an interview the new owner asked that her surname and hometown were not made public. She feared that she may become a target for groups that oppose the use of cloning techniques.

a) Explain why Little Nicky is exactly the same as Nicky. *(3 marks)*

b) Little Nicky's owner is concerned that she may become a target for anti-cloning groups.

   i) Give one example of a group or organisation that may disagree with the cloning of pet animals. *(1 mark)*

   ii) Why do you think such groups disagree with these cloning techniques? *(1 mark)*

4 a) Provide an explanation for the theory of evolution. *(2 marks)*

b) The text below provides a possible explanation for the evolution of the long legs of wading birds, by Lamarck.

When an animal undergoes change during their lifetime and then mates, they pass this change on to their offspring. For example, wading birds did not always have long legs. They gradually developed them after their ancestors started to feed on fish. As they walked into deeper water, they would stretch their legs to prevent their bodies from becoming wet, causing their legs to lengthen. Their new trait of longer legs would be passed on to their offspring, who would also stretch their legs. Over time, the legs of these wading birds became longer and longer.

Darwin would have provided a different explanation for the evolution of the long legs of these birds. What are the main differences between his explanation and that of Lamarck? *(3 marks)*

Chapter 5
How do rocks provide useful materials?

At the end of this chapter you should:

✓ appreciate how atoms, as the smallest particles in elements, can join together to form molecules in compounds;

✓ be able to use symbols and formulae to write balanced equations for chemical reactions;

✓ appreciate how rocks provide stone for building, metals and other useful materials;

✓ know how limestone is used to manufacture quicklime, slaked lime, cement, concrete and glass;

✓ understand how metals can be extracted from their ores;

✓ have considered the social, economic and environmental impact of quarrying, mining and extracting metals;

✓ have considered the benefits and drawbacks of using metals and recycling metals.

Figure 5.1 Iron ore, like that being mined in this photograph, is useless. You can't grow anything in it, you can't eat it and you can't build with it. But, if it is heated with limestone, coke and air, it produces iron, and from iron we can make steel, which is one of our most useful materials.

Figure 5.2 At one time, football boots were made of leather

❶ What raw material does leather come from?

❷ Today, football boots are made from polymers such as PVC and polyester. What raw material are the polymers made from?

Figure 5.3 Most of the outside of this building is glass

❸ a) Suggest two benefits of using glass for the building in Figure 5.3.
b) Suggest two drawbacks and risks.

5.1 What sorts of materials are there?

Materials that occur naturally, like iron ore, limestone, water and air are called **raw materials** or **naturally occurring materials**. Materials like iron and steel don't occur naturally, but we can make them using raw materials such as iron ore. Because of this, iron and steel can be called **manufactured materials**. These manufactured materials are useful products which we need for everyday modern life.

What are our most important raw materials?

Table 5.1 shows the five most important raw materials and some of the useful manufactured materials we can obtain from them.

The useful, manufactured materials are:
- either separated from the natural raw materials by processes such as distillation;
- or made from the raw materials by chemical reactions like iron from iron ore.

Raw material	Useful manufactured materials obtained from the raw material
Rocks	• Metals (iron, aluminium, copper) • Alloys (steel, brass) • Building materials (cement, glass)
Crude oil	• Fuels (petrol, diesel) • Plastics (polythene, PVC and polyester)
Air	• Nitrogen for making ammonia, nitric acid and fertilisers • Oxygen for breathing equipment
Seawater	• Table salt, sodium hydroxide, chlorine and hydrogen from brine (concentrated sodium chloride)
Plants	• Fruit and vegetables • Plant oils for cooking and medicines • Fuels from biomass materials

Table 5.1 The five most important raw materials

❹ Copy and complete the following table. The first line has been done for you.

Manufactured material	Which raw material in Table 5.1 did the manufactured material come from?
Metals in your T.V. Plastics in your ipod Ammonia in cleaning fluid Olive oil for cooking Polyester in your shirt/blouse Glass for bottles and jars	Rocks

Activity – Using the natural resources on Geecee

Geecee (Figure 5.4) is an imaginary island off the west coast of Britain. The island has strong winds from the west, high mountains with fast flowing rivers, thick forests along the west coast and peat bogs in the south east. There are no supplies of coal, oil or natural gas. At present no one lives on Geecee.

Suppose you have been sent to study whether people could live on the island and develop a fishing industry.

❶ What naturally occurring material would you use to build the first homes on Geecee? Explain your answer.
❷ Describe two ways of producing heat to cook food.
❸ Where would you build the port for the fishing fleet? Give two reasons for your choice of site.
❹ Where would you build homes on the island? Explain your answer.
❺ Which naturally occurring material(s) would you try to conserve? Explain your answer.

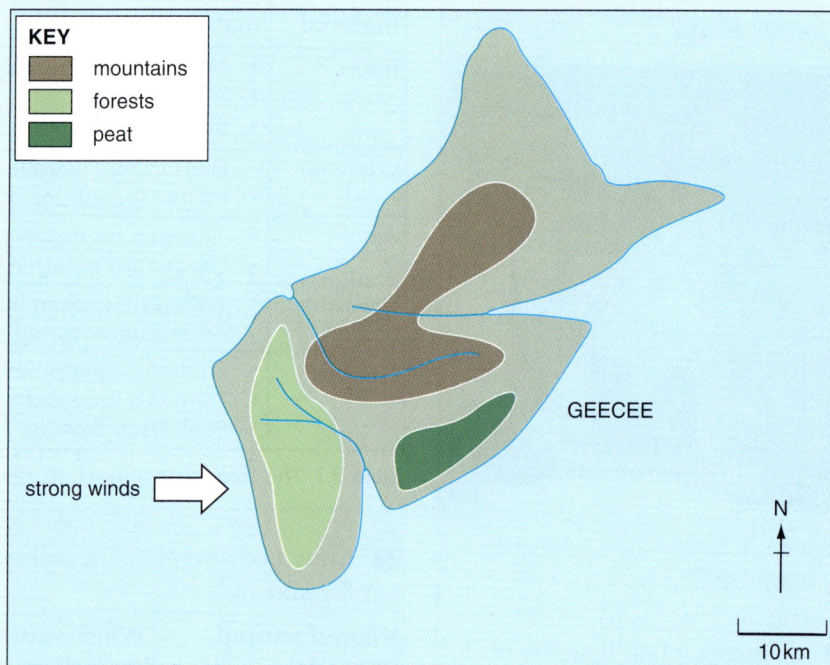

KEY
▮ mountains
▮ forests
▮ peat

strong winds ⇨

GEECEE

N

10km

Figure 5.4 The imaginary island of Geecee

5.2 Elements, compounds and mixtures

When electricity passes through molten sodium chloride, it breaks down to form sodium and chlorine (Figure 5.5). We can summarise the **chemical reaction** by writing a **word equation**.

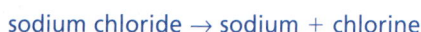

sodium chloride → sodium + chlorine

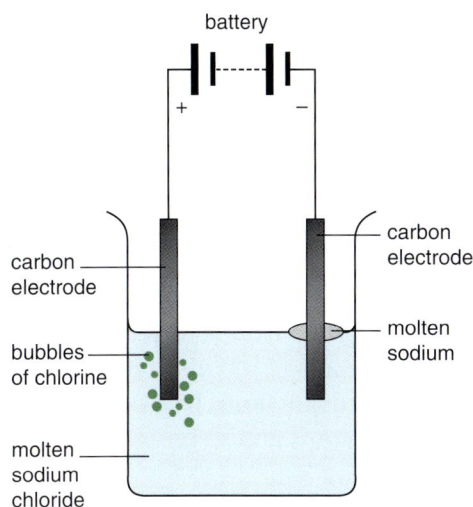

battery

carbon electrode

carbon electrode

bubbles of chlorine

molten sodium

molten sodium chloride

Figure 5.5 When electricity passes through molten sodium chloride, it breaks down into sodium and chlorine

No matter how the sodium and chlorine are treated, they cannnot be broken down into simpler substances.

Substances, like sodium and chlorine, that cannot be broken down into simpler substances are called **elements**. Elements are the simplest possible materials.

Other elements include aluminium, iron, oxygen and carbon. But substances like sodium chloride and water are not elements because they can be broken down into simpler substances. Substances, like sodium chloride and water, which contain two or more elements chemically joined together are called **compounds**.

There are about 100 different elements and, although there are millions and millions of different substances in the Universe, they all contain one or more of these elements.

For example, water is made of hydrogen and oxygen. Sand is made of silicon and oxygen and limestone contains calcium, carbon and oxygen. So, elements are the simplest building blocks for all substances.

Figure 5.6 shows the percentages of the five most common elements in the Earth's crust.

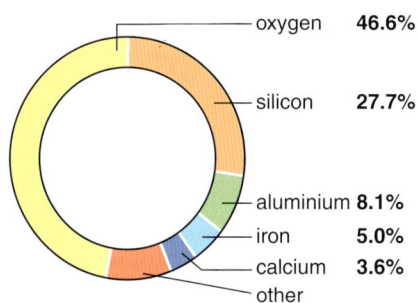

oxygen **46.6%**

silicon **27.7%**

aluminium **8.1%**

iron **5.0%**

calcium **3.6%**

other

Figure 5.6 The percentages of the five most common elements in the Earth's crust

An **element** is a substance containing only one kind of atom.

A **compound** contains two or more elements joined together chemically.

A **mixture** is two or more substances which are mixed together but *not* combined together chemically.

In an **alloy** other elements are mixed with metals to give the mixture particular properties.

5 a) What total percentage of the Earth's crust do these five elements make up?
b) What percentage of the Earth's crust do all the other elements make up?
c) What is the most abundant metal in the Earth's crust?
d) A student found five different sources of data for the percentage of sodium in the Earth's crust.
The values given were 2.7%, 3.5%, 2.6%, 2.9% and 3.0%.
Copy and complete the following sentences.

The **range in the data** is from the minimum value of _____% to the maximum value of _____%. One of the values given is very different to the other four. This **anomalous value** is _____%. Anomalous (unusual or irregular) values are usually ignored in calculating an average or mean value. When this is done:
The mean value for the % of sodium in the Earth's crust = _____%.

Most of the materials that occur naturally and that we use every day are *not* pure elements or pure compounds. They are **mixtures** of substances. They may be:
- mixtures of elements, such as **alloys**, which are metals mixed with other elements (for example, mild steel is mainly iron with about 0.2% carbon);
- mixtures of compounds, such as seawater, which contains salt (sodium chloride) and water.

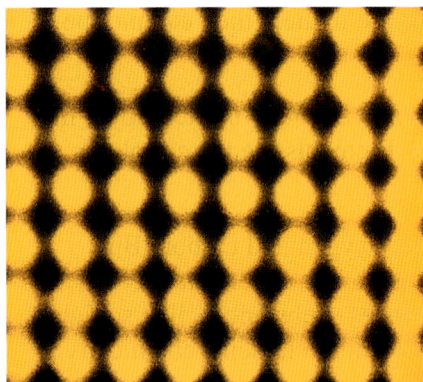

Figure 5.7 This photo of a gold crystal was taken through an electron microscope. Each yellow blob is a separate gold atom. The gold atoms are touching each other

An **atom** is the smallest particle of an element. All substances are made of atoms.

6 a) Estimate the diameter of one gold atom in the photo.
b) In order to estimate the diameter of a gold atom more accurately, it is better to measure a line of four or five of them and then divide by four or five. Why is this more reliable than measuring the diameter of just one gold atom? (Hint: Data is more reliable if you can be sure it is accurate.)
c) Calculate the actual diameter of a gold atom. (Assume the magnification is 40 000 000.)

A **reliable** measurement or result is one that can be repeated.

An **accurate** measurement is one that is close to the **true value**, which is the value that would be obtained if there were no errors in the measurement.

What are the particles in elements?

The smallest particles of an element are **atoms**.

Electron microscopes can magnify objects more than a million times. Using electron microscopes, it is possible to identify atoms. Figure 5.7 shows an electron microscope photo of gold.

As all substances are made of elements and all elements are made of atoms, it follows that all substances are made of atoms. Each element contains only one sort of atom. So, as there are about 100 different elements, there are also about 100 different kinds of atom. Iron contains only iron atoms, copper contains only copper atoms and so on.

Representing atoms with symbols

Atoms of each element can be represented by a chemical **symbol**. For example, O represents an atom of oxygen, Fe represents an atom of iron and C represents an atom of carbon. The symbol for an element can also be used as shorthand for the name of the element.

The names and symbols of all the common elements are shown in the Periodic Table on page 83.

7 Why is it useful to represent elements using symbols? (Hint: Suppose you had to write the word 'magnesium' many times.)

8 Use the Periodic Table containing symbols on page 83 to answer the following questions.
a) What are the symbols for carbon, calcium, cobalt, copper, chlorine and chromium?
b) What elements are represented by N, Ni, P, K, Si, Ag, S and Na?

Sometimes, it is useful to draw pictures of atoms as coloured circles with the symbol in the centre (Figure 5.8). This helps scientists to understand the structure of substances and how they react with each other.

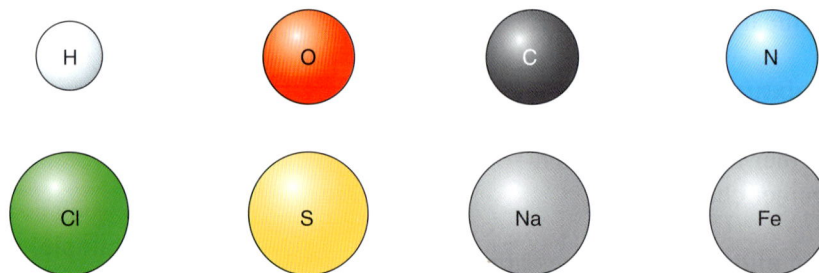

Figure 5.8 Pictures of atoms. Hydrogen atoms are usually shown as white circles, oxygen red, carbon black, nitrogen blue, chlorine green, sulfur yellow and metals grey.

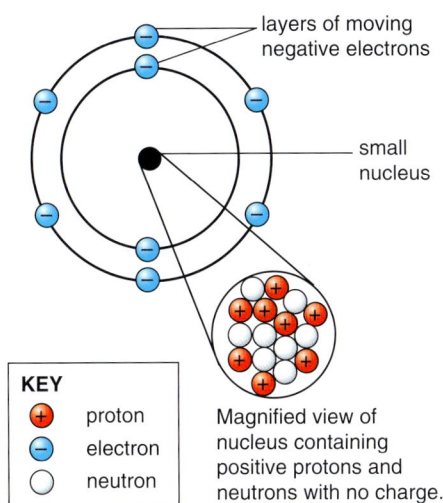

Figure 5.9 The nucleus and electrons in an atom of oxygen

KEY

⊕ proton
⊖ electron
○ neutron

layers of moving negative electrons

small nucleus

Magnified view of nucleus containing positive protons and neutrons with no charge.

How do atoms combine?

Atoms have a small positive **nucleus** surrounded by layers of moving negative **electrons**. The positive charge on the nucleus exactly cancels the negative charge on the electrons, so the overall charge on an atom is neutral.

The number of protons in the nucleus, called the atomic number or proton number, determines which element an atom is.

When elements react, their atoms join with other atoms to form compounds. This involves:
• either sharing electrons to form **molecules**;
• or transferring (giving and taking) electrons to form charged particles called **ions**.

When atoms of non-metals join together, they share electrons and form molecules.

In a molecule of water, two atoms of hydrogen combine with one atom of oxygen. The two H atoms and the one O atom are held together in **chemical bonds** by the attraction of their positive nuclei for shared electrons. Figure 5.10 shows how the electrons are shared in a molecule of water.

Atoms have a small central **nucleus** which has a positive charge.

Electrons are very small negatively charged particles that move around the nucleus.

A **molecule** is a particle containing two or more atoms joined by chemical bonds.

An **ion** is a charged particle formed from an atom by the loss or gain of one or more electrons.

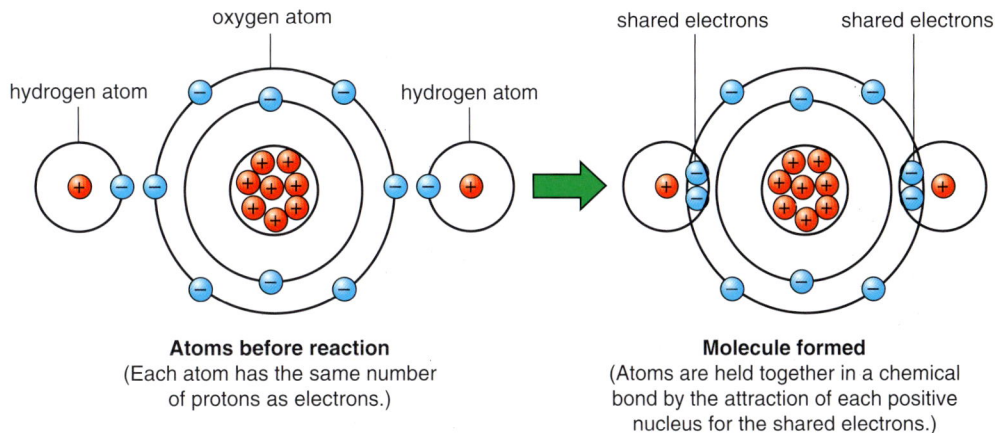

Figure 5.10 The formation of a molecule of water

Atoms before reaction (Each atom has the same number of protons as electrons.)

Molecule formed (Atoms are held together in a chemical bond by the attraction of each positive nucleus for the shared electrons.)

The symbols for elements can also be used to represent molecules in compounds. So, water is written as H_2O – two hydrogen atoms (H) and one oxygen atom (O). Carbon dioxide is written as CO_2 – one carbon atom (C) and two oxygen atoms (O).

The **formula** of a compound shows the number of atoms of the different elements joined together in one molecule.

'H_2O' and 'CO_2' are called molecular formulae, or just formulae for short.

These formulae show the number of atoms of the different elements joined together in one molecule of a compound.

9 How many atoms of the different elements are there in one molecule of
 a) ammonia;
 b) hydrogen chloride?

10 Look at Figure 5.11. What is the formula of: a) ammonia;
 b) hydrogen chloride?

11 a) Draw a picture for a molecule of methane (natural gas), CH_4. All the four hydrogen atoms are bonded to the carbon atom.
 b) What are the advantages of writing 'CH_4' for methane?

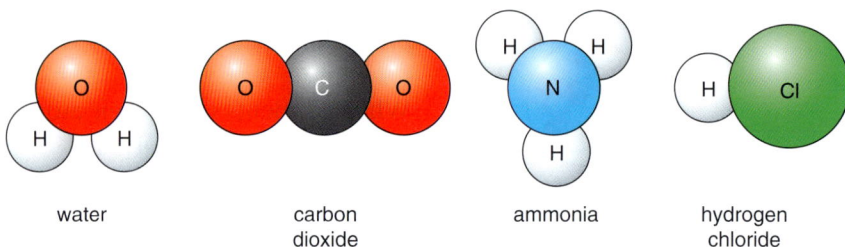

Figure 5.11 Molecules of water, carbon dioxide, ammonia and hydrogen chloride

When atoms of a metal join with atoms of a non-metal, they form ions.

So, compounds such as sodium chloride (NaCl) calcium oxide (CaO) and red iron oxide (Fe_2O_3) consist of ions, not molecules.

When these metal/non-metal compounds form (Figure 5.12), the metal atoms give up electrons to form positive ions and the non-metal atoms take electrons to form negative ions. In the crystals of these metal/non-metal compounds, there are lattices of positive and negative ions. The opposite charges hold the ions together in the lattices by strong chemical bonds.

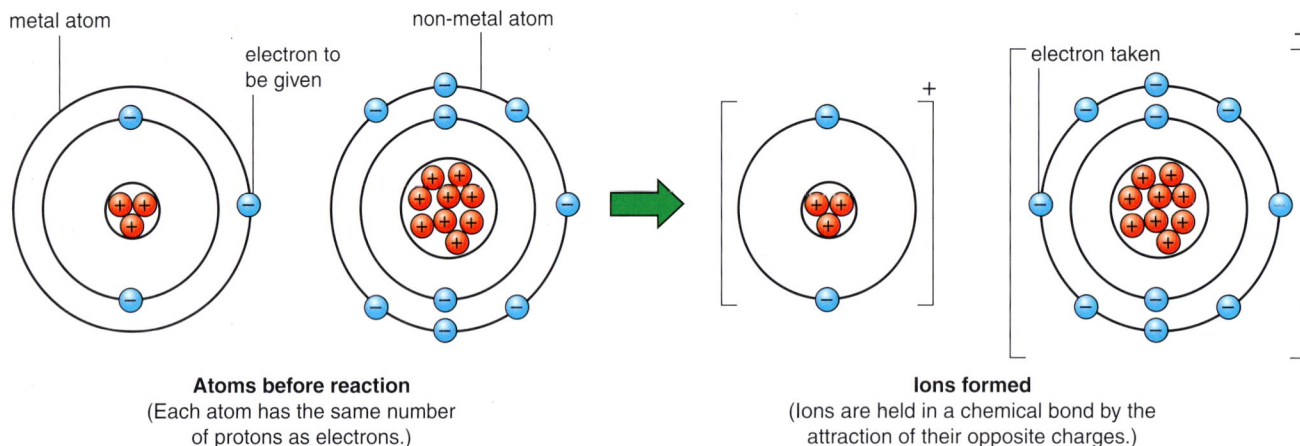

Atoms before reaction
(Each atom has the same number of protons as electrons.)

Ions formed
(Ions are held in a chemical bond by the attraction of their opposite charges.)

Figure 5.12 The formation of ions when atoms give and take electrons

an oxygen atom

Symbol O

Formula O_2

an oxygen molecule

Figure 5.13 An atom and a molecule of oxygen

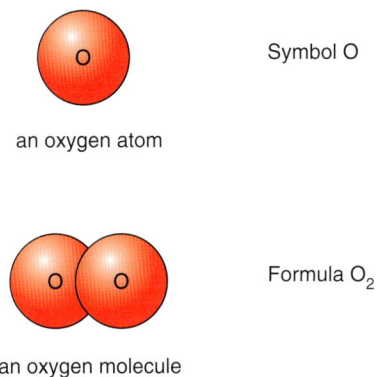

Atoms and molecules of elements

Almost all elements can be represented by their symbols. For example, Fe for iron, C for carbon. But, this is not the case with hydrogen, oxygen, nitrogen and chlorine. At normal temperatures, these elements exist as molecules containing two atoms joined together. So, hydrogen is best represented as H_2 not H, oxygen as O_2 not O, nitrogen as N_2 and chlorine as Cl_2 (Figure 5.13).

5.3 How are elements arranged in the Periodic Table?

In the **Periodic Table** (Figure 5.14) elements are arranged in order of their atomic number (the number of protons in an atom of the element).

So, the first element in the table is hydrogen (atomic number = 1), then helium (atomic number = 2), then lithium (atomic number = 3) and so on.

Figure 5.14 The Periodic Table (elements 58–71 and 90–103 have been omitted)

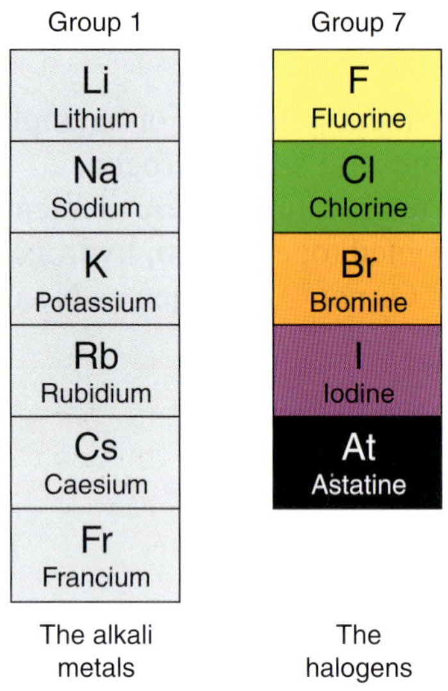

Group 1

Group 7

Li		F
Lithium		Fluorine
Na		Cl
Sodium		Chlorine
K		Br
Potassium		Bromine
Rb		I
Rubidium		Iodine
Cs		At
Caesium		Astatine
Fr		
Francium		

The alkali metals

The halogens

Figure 5.15 Group 1 and Group 7 of the Periodic Table

The Periodic Table is also set out so that elements with similar properties are in the same vertical column.

- The vertical columns are called **groups**. So, Group 1 contains the metals lithium, sodium and potassium, which have very similar properties, and Group 7 contains the non-metals chlorine and bromine which are also very similar (Figure 5.15).
- Some of the groups have names as well as numbers and these are shown below the group numbers across the top of Figure 5.14.
- The horizontal rows in the Periodic Table are called **periods**. So, period 1 contains just two elements – hydrogen and helium, and period 2 has eight elements from lithium (atomic number 3) to neon (atomic number 10).
- Metals are clearly separated from non-metals. The 20 or so non-metals are packed into the top right-hand corner, above the thick stepped line in Figure 5.14.
- **In each group of the Periodic Table the elements have similar properties**, but there is a gradual change in properties from the top to the bottom of the group.

The vertical columns in the Periodic Table are called **groups**. Elements in the same group have similar properties.

⓬ Look at Figure 5.15.
  a) Which of the elements, sodium or potassium, is the most reactive?
  b) Which element in Group 1 do you think is i) the most reactive; ii) the least reactive?
  c) Do the elements in Group 1 get more or less reactive as you go down the group from Li to Fr?

⓭ Draw a large outline of the Periodic Table similar to Figure 5.14. On your outline, indicate where you would find:
  a) metals;
  b) elements with atomic numbers 14 to 17;
  c) the noble gases;
  d) the transition metals;
  e) the most reactive metal;
  f) one magnetic element;
  g) two elements used in expensive jewellery;
  h) an element used to disinfect water supplies.

(5.4) ## Using symbols and formulae to write balanced equations

Figure 5.16 shows sparks from a sparkler. The sparks are tiny bits of burning magnesium. Let's use symbols and formulae to write an equation for this reaction.

When magnesium burns, it reacts with oxygen to form magnesium oxide.

$$\text{magnesium} + \text{oxygen} \rightarrow \text{magnesium oxide}$$

Chemists usually write symbols and formulae rather than names in equations. So, in the word equation, we should write Mg for magnesium, O_2 for oxygen and MgO for magnesium oxide.

$$Mg + O_2 \rightarrow MgO$$

This is more helpful than the word equation, *but it doesn't balance.* There are two oxygen atoms in O_2 on the left, but only one oxygen atom in MgO on the right. So MgO must be doubled to give

$$Mg + O_2 \rightarrow 2MgO$$

Unfortunately, *the equation still doesn't balance.* We now have one Mg atom on the left, but two Mg atoms in 2MgO on the right. This is easily corrected by writing 2Mg on the left to give:

$$2Mg + O_2 \rightarrow 2MgO$$

The numbers of different atoms are now the same on both sides of the arrow. This is a **balanced chemical equation**.

Figure 5.17 shows a picture equation for this reaction. Picture equations help us to understand how the atoms are rearranged in reactions.

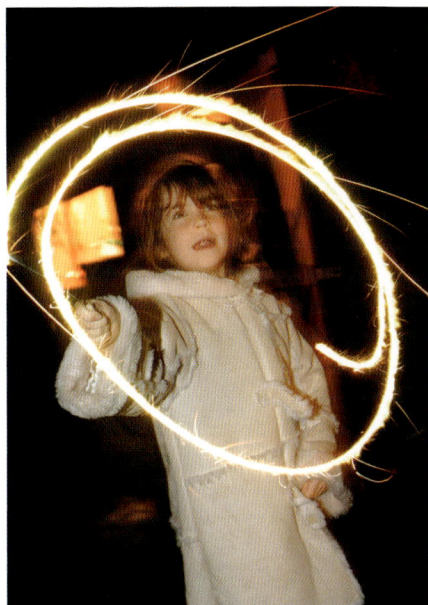

Figure 5.16 The sparks from a sparkler are tiny bits of burning magnesium

A **balanced chemical equation** shows the atoms involved in a chemical reaction. No atoms are lost or made in a chemical reaction, so the numbers of each kind of atom are the same on both sides of a balanced equation.

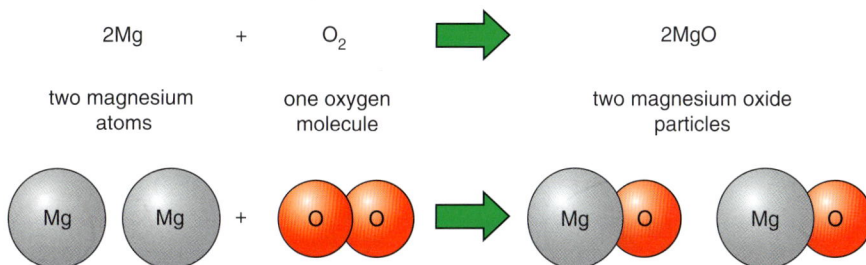

Figure 5.17 A picture equation for the reaction between magnesium and oxygen

Picture equations show that no atoms are lost during a chemical reaction. The atoms in the reactants are all still there at the end of the reaction but they are arranged differently to make different substances. As all the atoms are present at the start and finish of the reaction, the mass of the products must equal the mass of the reactants.

This is summarised in **the law of conservation of mass,** which says: in any chemical change, the total mass of the products equals the total mass of the reactants.

No atoms are lost or made during chemical reactions so we can write balanced equations by making sure there are the same number of atoms of each element on both sides of the arrow.

The following example shows the three steps to follow.

Step 1 Write a word equation.

hydrogen + oxygen → water

Step 2 Write symbols or formulae for the reactants and products.

$$H_2 + O_2 \rightarrow H_2O$$

Remember that hydrogen, oxygen, nitrogen and chlorine exist as molecules and are written as H_2, O_2, N_2 and Cl_2. All other elements are shown as single atoms (C for carbon, Fe for iron).

Step 3 Balance the equation by making the number of atoms of each element the same on both sides.

$$2H_2 + O_2 \rightarrow 2H_2O$$

Figure 5.18 shows a picture equation for this reaction.

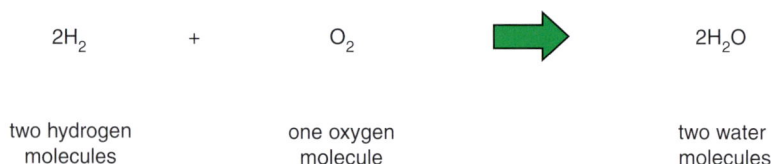

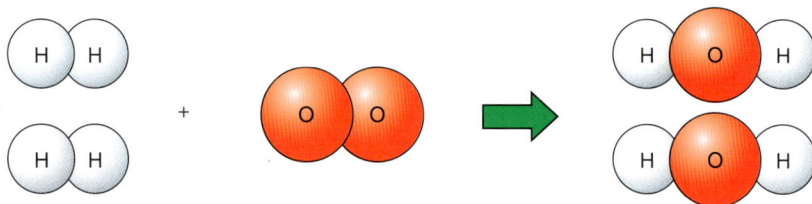

$2H_2$	+	O_2		$2H_2O$
two hydrogen molecules		one oxygen molecule		two water molecules

Figure 5.18 A picture equation for the reaction of hydrogen with oxygen to form water

Remember that you must never change a formula to make an equation balance. The formula for water is always H_2O and never HO or HO_2. Similarly, the formula of magnesium oxide is always MgO and never MgO_2 or Mg_2O.

You can only balance an equation by putting numbers in front of symbols or formulae, for example, 2Mg, 2MgO, $2H_2$ and $2H_2O$.

Figure 5.19 In a barbecue, charcoal (carbon) burns in oxygen in the air to form carbon dioxide

14 a) Write a word equation for the reaction of charcoal (carbon) burning in oxygen to form carbon dioxide.
    b) Why is it important that equations balance?
    c) Why is the formula for water always H_2O and never HO or HO_2?

Balanced equations are more useful than word equations because they show:
- the symbols and formulae of the reactants and products;
- the relative numbers of atoms and molecules of the reactants and products;
- the rearrangement of atoms from reactants to products.

State symbols

State symbols are used in equations to show the state of a substance. (s) after a formula or symbol indicates that the substance is a solid. (l) is used for a liquid, (g) for a gas and (aq) for an aqueous solution (where a substance is dissolved in water). For example,

$$2Mg(s) + O_2(g) \rightarrow 2MgO(s)$$

Figure 5.20 When natural gas burns on a hob, methane (CH_4), reacts with oxygen in the air to form carbon dioxide and water

15 Copy and balance the equation for the reaction in Figure 5.20.

$$CH_4 + \underline{} O_2 \rightarrow CO_2 + \underline{}H_2O$$

16 Propane, C_3H_8, is used in large red cylinders as a portable fuel in caravans and in some homes. The word equation and balanced chemical equation for burning propane are shown below.

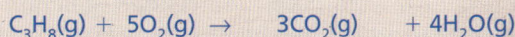

propane + oxygen → carbon dioxide + water

$$C_3H_8(g) + 5O_2(g) \rightarrow 3CO_2(g) + 4H_2O(g)$$

Write down six things that the balanced equation tells you and which you could not know from the word equation. This should help you to appreciate why chemists like to write balanced equations for chemical reactions.

17 a) The following equations are not balanced. Write out the equations and balance them.
        i) $N_2 + H_2 \rightarrow NH_3$
        ii) $Na + H_2O \rightarrow NaOH + H_2$
        iii) $C_2H_6 + O_2 \rightarrow CO_2 + H_2O$
    b) Write balanced equations with state symbols for the following word equations.
        i)   calcium + oxygen → calcium oxide (CaO)
        ii) hydrogen + chlorine → hydrogen chloride
        iii) zinc + hydrochloric acid (HCl) → zinc chloride ($ZnCl_2$) + hydrogen

5.5 How do rocks provide building materials?

Rocks can be quarried or mined to provide essential building materials such as stone for building homes, factories and offices. One of the most important rocks and naturally occurring resources is limestone. Limestone is mainly calcium carbonate, $CaCO_3$. It contains calcium ions, Ca^{2+} combined with carbonate ions, CO_3^{2-}. Each carbonate ion has one carbon atom bonded to three oxygen atoms with an overall charge of $2-$. Limestone is quarried and used as building stone. The stone can be broken into smaller pieces and used as aggregate and chippings in concrete. Large amounts of limestone aggregate are used for making roads every year.

Figure 5.21 Blocks of limestone have been used to build castles, cathedrals and houses for hundreds of years

Figure 5.22 Mining engineers like those in these photos must survey quarry sites with great care before setting explosives to dislodge the rock and quarry it safely

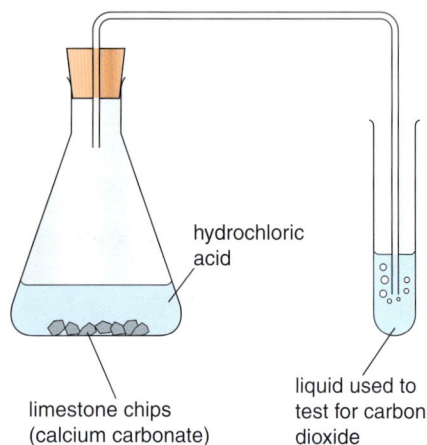

Using limestone to neutralise acidity

Limestone chippings are crushed to produce powdered limestone. This is used to neutralise acidity in soils and lakes. In this reaction, the limestone (calcium carbonate) reacts with acid to produce a calcium compound, water and carbon dioxide, which causes the mixture to 'fizz'. For example, with hydrochloric acid:

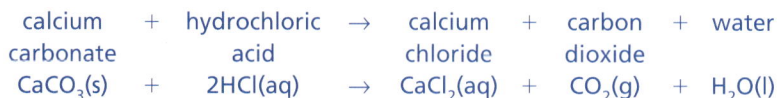

$$\text{calcium carbonate} + \text{hydrochloric acid} \rightarrow \text{calcium chloride} + \text{carbon dioxide} + \text{water}$$

$$CaCO_3(s) + 2HCl(aq) \rightarrow CaCl_2(aq) + CO_2(g) + H_2O(l)$$

Figure 5.23 Calcium carbonate reacting with hydrochloric acid, with a test for the carbon dioxide produced

Figure 5.23 shows calcium carbonate and hydrochloric acid reacting with a test for the carbon dioxide produced.

Decomposing limestone to make quicklime and slaked lime

Limestone (calcium carbonate) decomposes when it is heated strongly. The products are calcium oxide (commonly called quicklime) and carbon dioxide.

$$\text{calcium carbonate} \rightarrow \text{calcium oxide} + \text{carbon dioxide}$$

$$CaCO_3(s) \rightarrow CaO(s) + CO_2(g)$$

18 a) Name the liquid used to test for carbon dioxide.
b) What happens to this liquid as carbon dioxide bubbles into it?
c) Why is the soil in limestone areas neutral or slightly alkaline?

This is an example of **thermal decomposition** – using heat to break down a compound.

Carbonates of other metals decompose on heating in a similar way to calcium carbonate. The lower a metal is in the reactivity series, the more easily its carbonate decomposes.

Calcium oxide (quicklime) reacts vigorously with water to form calcium hydroxide (often called slaked lime). The reaction is strongly exothermic, which means that heat is produced.

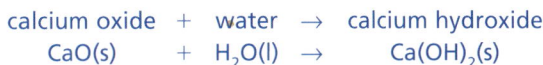

$$\text{calcium oxide} \ + \ \text{water} \ \rightarrow \ \text{calcium hydroxide}$$
$$CaO(s) \ \ + \ H_2O(l) \ \rightarrow \ \ Ca(OH)_2(s)$$

Both quicklime and slaked lime are useful substances. They are used in industry as cheap alkalis to neutralise acidity. Calcium hydroxide (slaked lime) is used by water companies to neutralise acid in water supplies and to make bleaching powder. Farmers and gardeners also use calcium hydroxide like powdered limestone on acid soils to neutralise acidity.

Figure 5.24 summarises the reactions of calcium carbonate, calcium oxide and calcium hydroxide.

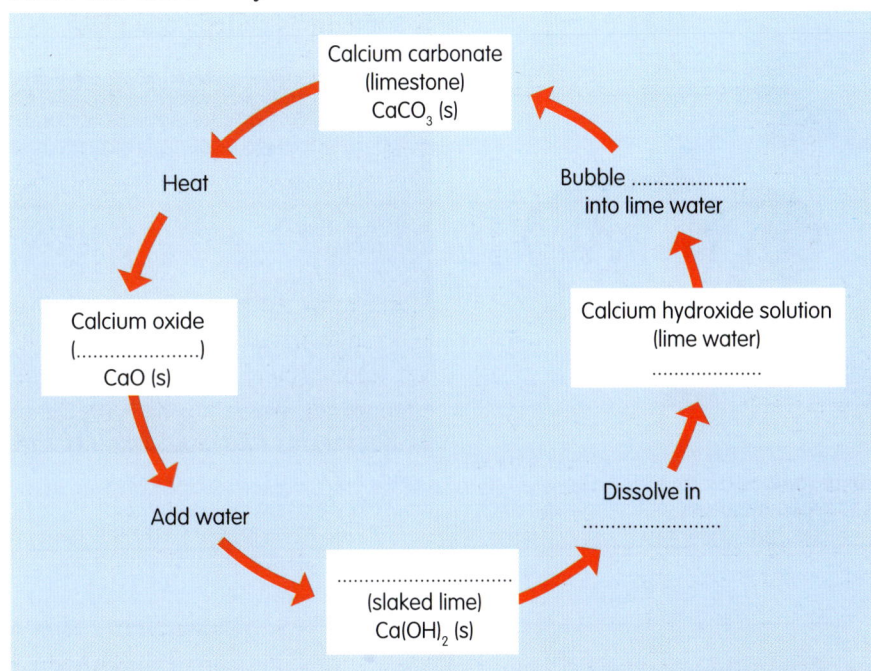

> Calcium hydroxide is slightly soluble in water. The solution is called lime water, which can be used to test for carbon dioxide. When carbon dioxide is bubbled into lime water, calcium carbonate forms as a milky precipitate.

19 a) Limestone is insoluble in water, but slaked lime (calcium hydroxide) is slightly soluble. Why does this make slaked lime better than limestone for use on acid soils?
   b) Write a balanced equation for the action of heat on copper carbonate ($CuCO_3$).

20 Copy Figure 5.24 and fill in the blank spaces.

Figure 5.24 The reactions of calcium carbonate, calcium oxide and calcium hydroxide

How is limestone used to manufacture other useful materials?

In addition to quicklime and slaked lime, limestone also provides a starting point for the manufacture of cement, concrete and glass. It is also used in the extraction of iron from iron ore (see Section 5.7). All these uses make limestone an extremely valuable resource for the chemical and building industries (Figure 5.25).

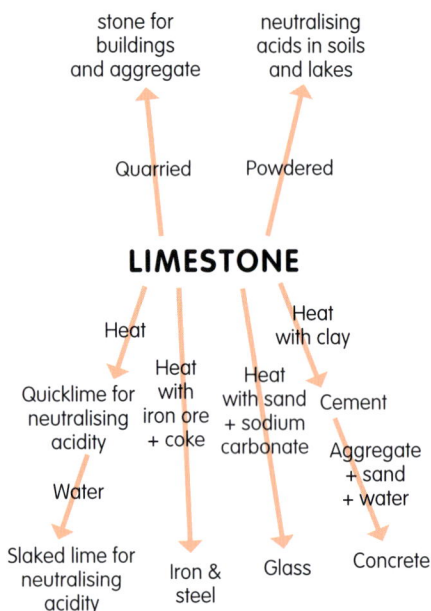

stone for buildings and aggregate

neutralising acids in soils and lakes

Quarried

Powdered

LIMESTONE

Heat

Heat with clay

Heat with iron ore + coke

Heat with sand + sodium carbonate

Cement

Quicklime for neutralising acidity

Water

Aggregate + sand + water

Slaked lime for neutralising acidity

Iron & steel

Glass

Concrete

Figure 5.25 Important uses and products of limestone

Cement is made by heating limestone with clay in a kiln. When cement is used, it is normally mixed with two or three times as much sand as well as water. This mixture is called **mortar**. Mortar reacts slowly and sets to form a very hard material. Bricklayers use mortar to hold bricks firmly together.

Concrete is made by mixing mortar with aggregate (small pieces of broken rock). As the mortar sets around the aggregate, it produces a hard, stone-like building material. The mixture will even set under water at room temperature. Isn't that strange and extraordinary?

Glass is usually made by heating a mixture of metal oxides or metal carbonates with pure sand (silicon dioxide, SiO_2) in a furnace. At the high temperatures in the furnace, carbonates decompose to oxides and bubbles of carbon dioxide escape from the mixture. If the mixture is heated further, a runny liquid forms. This liquid is allowed to cool until it is thick enough to be moulded or blown into different shapes. On further cooling, the glass sets solid.

Ordinary glass for bottles and windows is made by heating a mixture of limestone (calcium carbonate), soda (sodium carbonate) and sand. This is sometimes called soda glass.

Figure 5.26 Cutglass dishes and ornaments are made of lead glass, which is harder and shinier than ordinary glass

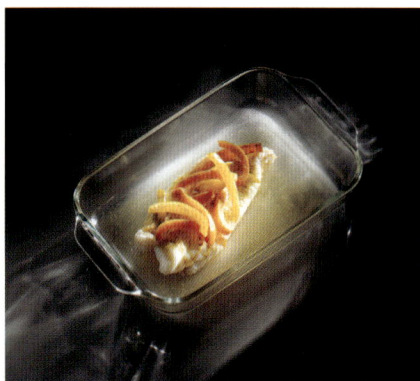

Figure 5.27 Glass ovenware and laboratory glassware are made of borosilicate glass (Pyrex®), which is heat resistant

Figure 5.28 Blue glass is made by adding cobalt oxide to the usual constituents for making glass

㉑ Various types of special glass are produced for particular purposes. Some of these are shown in Figures 5.26, 5.27 and 5.28.

a) What substances do you think are added to the usual constituents in order to make
   i) lead glass; ii) Pyrex®?
b) Which element produces the colour in blue glass?

c) Which other elements might produce colour in glass?
d) What benefits do these photos show in developments in the use of glass?
e) What are the risks and drawbacks of using glass?

Activity – Quarrying or countryside?

There are important environmental, social and economic issues involved in the quarrying and mining of rocks and ores. These issues are well illustrated in the UK by the quarrying of limestone and the production of building materials from it. Limestone occurs in some of the most beautiful areas of Britain: the Yorkshire Dales, the Peak District in Derbyshire, the Chilterns in Buckinghamshire and the Sussex Downs. The quarrying of limestone can spoil the countryside and create environmental problems.

On the other hand, limestone is a very important raw material for industry. Every year about 90 million tonnes of limestone are quarried in Britain. The limestone industry provides useful products for society; it also creates jobs and increases our wealth as a country.

So, how do we balance the benefits of quarrying with the problems it causes?

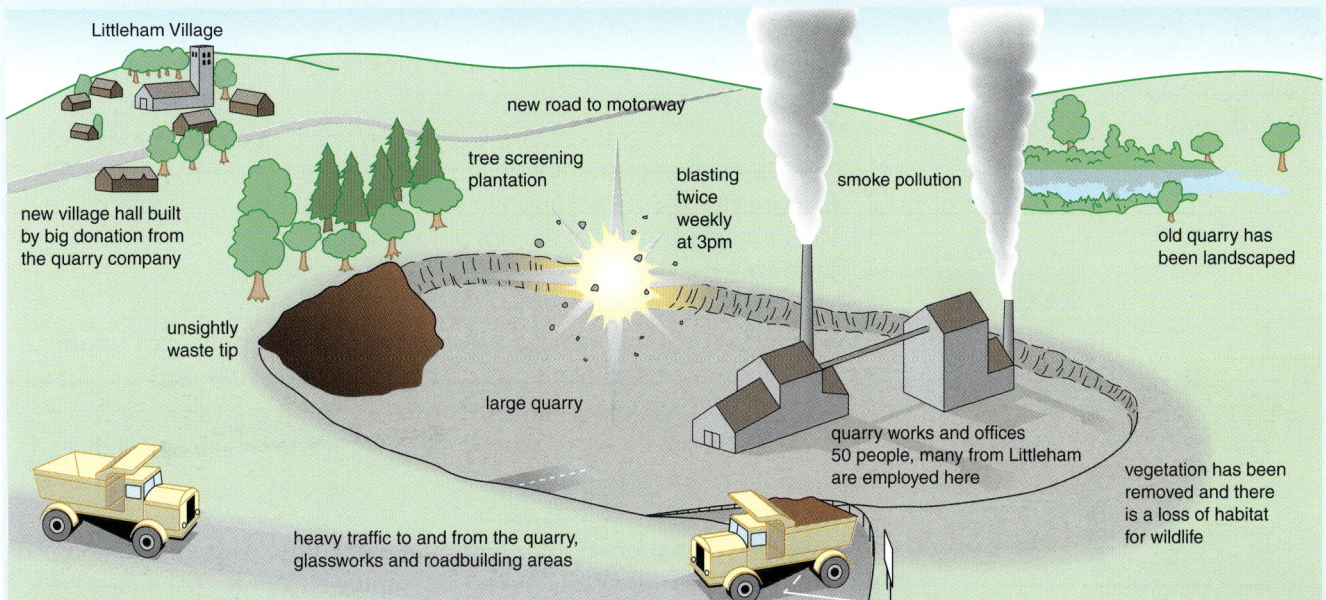

Figure 5.29 The quarry and surrounding area near Littleham

Look carefully at Figure 5.29, which shows a quarry near Littleham.

❶ Suppose you live in Littleham. Make separate lists showing three advantages and three disadvantages of the nearby quarry for Littleham.

❷ Suppose you are the Chairman of the Littleham Residents' Association. Write a letter to the Chief Executive of Limestone UK, the quarry operators, expressing the complaints you have had from members of the Residents' Association.

❸ Suppose you are the Chief Executive of Limestone UK. Write a reply to the Chairman of the Littleham Residents' Association describing your efforts to reduce problems caused by the quarry and the improvements and advantages it has created for the area and people of Littleham.

❹ Science can help us in many ways, but there are some questions that science cannot answer at all. These tend to be questions where beliefs are important, where views are personal or where we cannot obtain reliable evidence.
Science cannot tell us whether the quarry at Littleham has provided the residents with a richer or a poorer lifestyle. Why not?

Figure 5.30 Attractive crystals of gold on quartz

An **ore** is a rock or mineral from which a metal can be extracted.

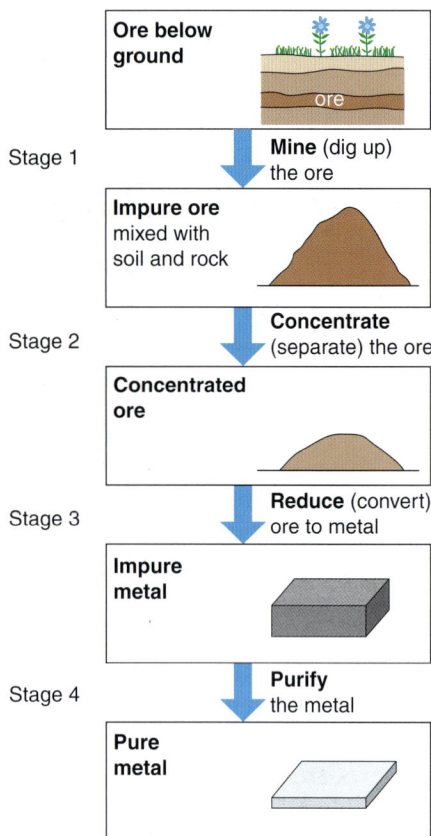

Figure 5.31 The four stages in extracting a metal from its ore

5.6 How do rocks provide metals?

Besides building materials, rocks also provide important metal **ores**. These ores contain enough metal or metal compounds to make it economic to extract the metal (Table 5.2).

Name of ore	Name and formula of metal compound in the ore	Metal obtained
Bauxite	Aluminium oxide, Al_2O_3	Aluminium
Iron ore (haematite)	Red iron oxide, Fe_2O_3	Iron
Copper pyrites	Copper sulfide, CuS, and iron sulfide, FeS	Copper

Table 5.2 The ores from which we extract some important metals

From ores to metals

Extracting (getting) metals from their ores usually involves four stages (Figure 5.31):

1 Mining (digging up) the ore.
2 Concentrating (separating) the ore.
3 Reducing (converting) the ore to the metal.
4 Purifying the metal.

Mining metal ores involves quarrying, tunnelling or open-cast mining. After mining, the ore must be separated from impurities such as soil and waste rock. This is called **concentrating** the ore. First the rock is crushed. Then, the ore is separated from the waste by a process that relies on their different densities.

The ores of some metals are in very limited supply. Others are more plentiful, but even the richest ores are impure. Iron ore (haematite) is over 80% pure Fe_2O_3 in many parts of the world, but copper ores rarely contain more than 1% of the pure copper compound.

A few metals, such as gold, are so unreactive that they are found in the Earth as the metals themselves. So, extracting these metals does not involve a chemical process. In the case of metals like gold, the concentrated ore is the impure metal itself. So, these metals can be obtained by simply mining, concentrating and purifying the metal.

Most metals are too reactive to exist on their own in the Earth. Their ores are compounds – usually metal oxides or substances such as sulfides and carbonates that can easily be changed into oxides. The metal can then be obtained from the metal oxide by removing oxygen. This loss of oxygen by the metal oxide is an example of reduction. Reduction is studied in more detail in Section 5.7.

Sodium — Most reactive
Calcium
Magnesium
Aluminium
Carbon
Zinc
Iron
Lead
Copper
Silver
Gold — Least reactive

Decreasing reactivity

Figure 5.32 The reactivity series showing the position of carbon

There are two main methods of reducing metal compounds depending on the position of the metal in the reactivity series (Figure 5.32).

Notice that carbon has been included in Figure 5.32 even though it is *not* a metal. Any element higher in the reactivity series can displace an element lower down from its compounds.

Figure 5.32 shows that metals, like zinc and iron, in the middle of the reactivity series are less reactive than carbon. This means that they can be extracted by reduction of their oxides with carbon (coke) or carbon monoxide.

22 a) Copy and complete the following word equation for the formation of tin.
tin oxide + carbon → ?
b) Now write a balanced chemical equation for the process.
c) Why do you think that tin is no longer produced in Cornwall?

23 a) Name two metals, other than zinc or iron, that can be obtained by reducing their oxides with carbon (coke) or carbon monoxide.
b) Name two metals, other than sodium and aluminium, that are obtained by electrolysis of their molten compounds.

Figure 5.33 Derelict tin mine buildings in Cornwall. Tinstone (tin oxide, SnO_2) was once mined in Cornwall and reduced to tin using carbon (coke) in furnaces near the mines. Decisions about whether to mine metal ores are often made on economic grounds. If the price of tin increased greatly in world markets, tin mining might well return to Cornwall

Metals like sodium and aluminium are above carbon in the reactivity series, so their compounds cannot be reduced to the metals using carbon. They are extracted by **electrolysis**. This involves decomposing the molten oxide or chloride to the metal using electricity.

Counting the cost of extracting metals

Extracting metals from their ores involves turning huge quantities of raw materials into useful and much more valuable metals. The metals can be used to manufacture a vast range of desirable products – vehicles, tools, pans, cutlery, cans, jewellery, pipes and girders.

jobs with good salaries

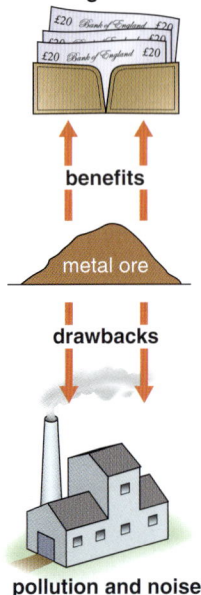

benefits

metal ore

drawbacks

pollution and noise

Figure 5.34 One of the benefits and one of the drawbacks of exploiting metal ores

24 Look carefully through the last sub-section headed 'Counting the cost of extracting metals'.
a) Make a list of other benefits and drawbacks similar to the two already shown in Figure 5.34.
b) Sketch out Figure 5.34 and add further artwork and labels to your own diagram to show benefits and drawbacks.

The extraction of metals followed by the production and sale of valuable metal products creates jobs for many people. This improves their standard of living and adds to the wealth of a nation. But we, as a society, don't get these benefits for nothing.

Social costs

People who work in the mining and metals industries are exposed to health and safety risks. Some of the processes involve chemicals which are toxic (poisonous) and workers can suffer damage to their hearing from loud factory noise. It is important to remember, however, that it is possible to work safely with hazardous chemicals and in noisy factories by taking suitable precautions. Adverse effects may extend beyond the mines and factories to people living in the area. But this will not happen if appropriate health and safety regulations are followed.

Large industrial operations like mining and quarrying often involve rapid changes in the number of people employed and living in an area. This can put great strain on social services like schools and hospitals.

Environmental costs

Where there is mining, quarrying and the use of heavy machinery, wildlife habitats and farmland will be destroyed. The extraction of ores and transport of materials also create noise and pollution. The pollution comes in various forms: air pollution from factory chimneys and vehicle exhausts, dust from blasting and unsightly tips of waste materials.

Economic costs

Large industrial operations also incur huge economic costs due to the use of expensive machinery and materials, particularly fuels. Fuels are needed to operate machinery, to heat buildings and maintain chemical processes, as well as to transport workers and materials.

When you next buy a can of Coke® or a piece of jewellery remember that although these items may add to your enjoyment, their production has social, environmental and economic costs.

5.7 How is iron extracted from iron ore?

The main raw material for making iron is iron ore (haematite). Haematite is impure iron oxide, Fe_2O_3. The iron ore is usually obtained by open-cast mining.

Iron is extracted from the iron ore in a **blast furnace**. A blast furnace is a large tower about 15 metres tall. Figure 5.35 on the next page shows a diagram of a blast furnace with an explanation of the chemical processes alongside.

① Solid raw materials (iron ore, coke (carbon) and limestone) are added at the top of the furnace.

② Blasts of hot air (which give the furnace its name) are blown in near the bottom of the furnace.

③ Oxygen in the blasts of air causes the coke (carbon) to burn, forming carbon dioxide and releasing energy (heat).

$$\text{carbon} + \text{oxygen} \rightarrow \text{carbon dioxide}$$
$$C + O_2 \rightarrow CO_2$$

④ At the high temperatures in the furnace, carbon dioxide reacts with more coke (carbon) to form carbon monoxide.

$$\text{carbon dioxide} + \text{carbon} \rightarrow \text{carbon monoxide}$$
$$CO_2 + C \rightarrow 2CO$$

⑤ The carbon monoxide reacts with the iron ore (red iron oxide) producing carbon dioxide and molten iron.

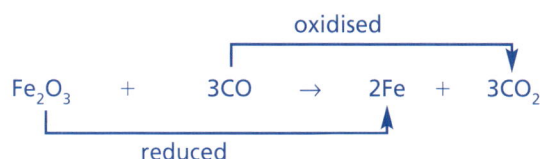

red iron oxide + carbon monoxide → iron + carbon dioxide

oxidised

$$Fe_2O_3 + 3CO \rightarrow 2Fe + 3CO_2$$

reduced

⑥ Molten iron runs to the bottom of the furnace and is tapped off from time to time.

Labels on figure: skip; gas outlet; load of iron ore, coke and limestone; brick lining to furnace; 1000°C; 1500°C; 2000°C; blasts of hot air; molten slag (impurities); outlet for slag; outlet for iron; molten iron

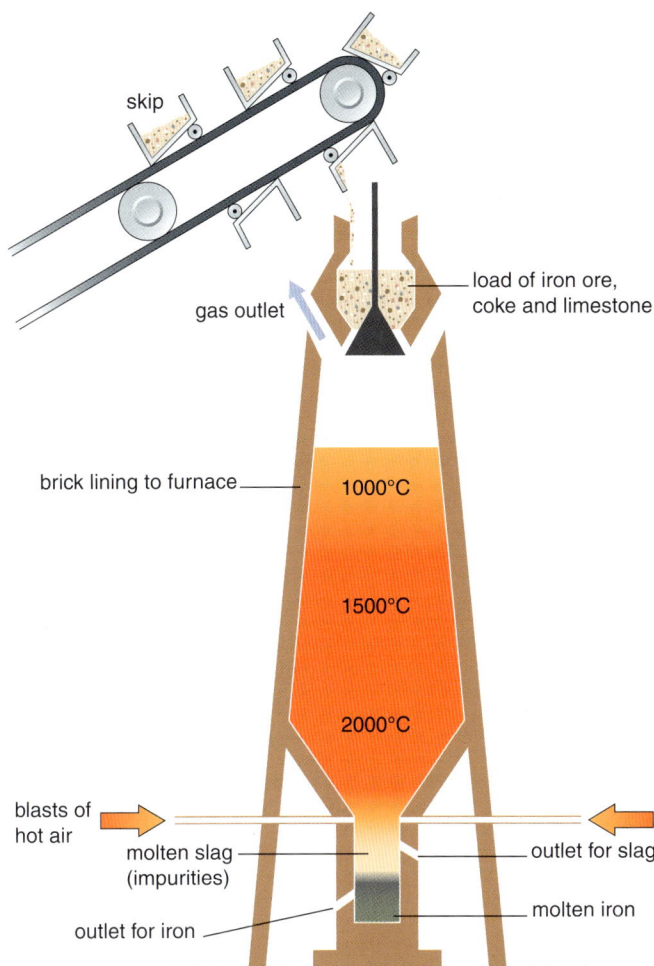

Figure 5.35 Extracting iron from iron ore in a blast furnace

㉕ The reaction between carbon dioxide and coke (carbon) to form carbon monoxide is a redox reaction.
a) What is a redox reaction?
b) In the above redox reaction, which substance is:
  i) oxidised;
  ii) reduced;
  iii) the oxidising agent;
  iv) the reducing agent?

Oxidation is the addition of oxygen to a substance.

In the reaction between Fe_2O_3 and CO, carbon monoxide gains oxygen forming carbon dioxide. This gain of oxygen is called **oxidation** and the carbon monoxide is said to be oxidised. At the same time, iron oxide loses oxygen. This loss of oxygen is called **reduction** and the iron oxide is said to be reduced. Iron oxide, which supplies oxygen, is described as the oxidising agent and carbon monoxide, which takes oxygen, is described as the reducing agent.

Oxidation and reduction always occur together. If one substance gains oxygen and is oxidised, another substance must lose oxygen and be reduced. We call the combined process **redox** (**red**uction + **ox**idation).

It may seem confusing, but notice that during redox reactions:
• the oxidising agent (in this case iron oxide) is reduced;
• the reducing agent (in this case carbon monoxide) is oxidised.

Why is limestone used in the furnace?

The main impurity in iron ore is sand (impure silicon dioxide, SiO_2). This is removed by limestone.

At the high temperatures in the furnace, limestone decomposes forming calcium oxide and carbon dioxide. The calcium oxide, which is basic, reacts with sand (SiO_2), which is acidic, to form 'slag', calcium silicate.

$$\text{calcium oxide} + \text{silicon dioxide} \rightarrow \text{calcium silicate}$$
$$CaO + SiO_2 \rightarrow CaSiO_3$$

The molten 'slag' falls to the bottom of the furnace and floats on the molten iron. This can be tapped off at a different level from the molten iron. The 'slag' is used in road making and cement manufacture.

Why is iron converted to steel?

Iron from the blast furnace contains about 96% iron. The main impurity in this iron is carbon. This makes it brittle, so it has only limited uses. Removing all the impurities from the iron would produce pure iron. This is too malleable (easily shaped) and too soft for most uses. Most hot molten iron from the blast furnace goes straight to a steel making furnace. Here it is converted into steels with the ideal strength and hardness. Steels are alloys – mixtures of iron with carbon and often other metals.

Steel is made by blowing oxygen under pressure onto the hot, molten impure iron. The oxygen converts excess carbon to carbon dioxide which escapes as a gas.

Using alloys

Alloys can be designed and manufactured to have properties for specific uses. Some are designed for hardness, some for resistance to corrosion and others have special magnetic or electrical properties.

Alloys are usually made by melting the main metal and then dissolving the other elements in it.

The most important alloys are steels. The composition, properties and uses of various steels are shown in Table 5.3.

Figure 5.36 Molten iron being poured from a furnace

Type of steel	Composition	Properties	Uses
Low-carbon steel (mild steel)	99.8% iron 0.2% carbon	Easily pressed into shapes	Car bodies
High-carbon steel	98.0% iron 1.7% carbon 0.3% manganese	Hard but brittle	Tools
Stainless steel	73.7% iron 0.3% carbon 18.0% chromium 8.0% nickel	Hard and resistant to corrosion	Cutlery, pans

Table 5.3 The composition, properties and uses of various steels

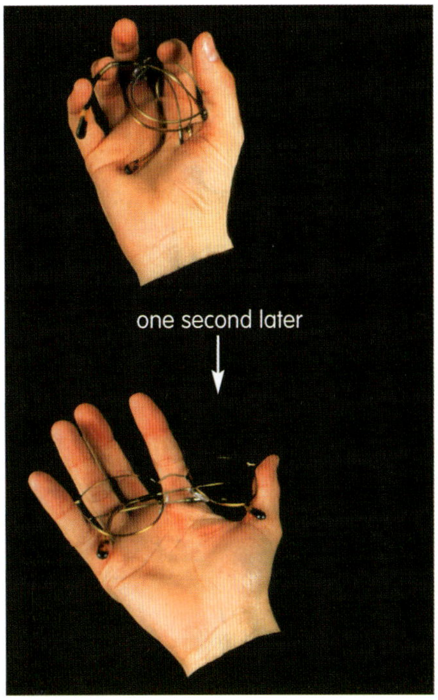

one second later

Figure 5.37 Spectacle frames made of 'smart alloys' will revert to their original shape in a second, after being bent and twisted

Most metals in everyday use are alloys. Like iron, pure copper, pure aluminium and pure gold are too soft for most uses. So, they are mixed with small amounts of other metals to make them harder. During the last 40 years, aluminium alloys have been used more and more. These include duralumin, which contains 4% copper. Aluminium alloys are light, strong and corrosion resistant. They are used for aircraft bodywork, overhead electricity cables and lightweight tubing.

In recent years, **smart alloys**, sometimes called 'memory metals', have been developed. These can return to their original shape after being deformed. Smart alloys are excellent for use in spectacle frames and in the braces fitted by dentists. The smart alloy braces are made so that, after fitting, they will return to their original shape and pull the teeth into better alignment.

The main elements in important alloys are transition metals in the central block of the Periodic Table. These transition metals include iron, copper, chromium, nickel and titanium.

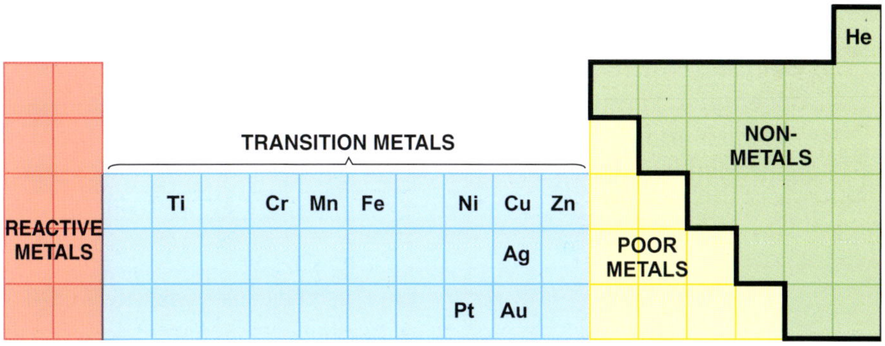

Figure 5.38 The position of the transition metals in the Periodic Table

Like other metals, transition metals are good conductors of heat and electricity. They can also support heavy loads and can be bent or hammered into shape. So, they are useful as structural materials and for making things that must conduct heat or electricity easily. Steel is one of our most important structural materials. It is used in girders, joists and bridges. Copper has properties that make it useful for electrical wiring and in pipes for plumbing.

Aluminium and titanium are two other useful metals because of their low density and resistance to corrosion. Both metals occur in ores as their oxides, but these oxides cannot be extracted by reduction with carbon. Current methods of extracting the two metals are expensive because large amounts of energy are needed and there are several stages in each process. These costs have limited the use of titanium.

Figure 5.39 A hip joint made of titanium alloy

26 This question is about new ways to extract copper. Answer the questions as you read the passage.

Copper ores contain copper sulfide (CuS). At one time, all copper was extracted from these sulfide ores by first converting the copper sulfide to copper oxide and then reducing this to copper by heating with carbon.

During the last 20 years, the supply of copper-rich ores has become very limited. This has led chemists to look for new ways of extracting copper from low-grade ores.

a) What is the main environmental problem of mining low-grade ores that contain vast amounts of worthless rock?

Fortunately, there are helpful bacteria in most copper ores. These bacteria use oxygen in the air to oxidise insoluble copper sulfide to soluble copper sulfate ($CuSO_4$).

b) Write a word equation for this reaction.
c) Use your word equation to write a balanced chemical equation with state symbols.

The bacteria found in the ore can tolerate:
• acidic conditions;
• heat generated by the reaction;
• the copper compounds which are poisonous to most organisms.

To extract the copper a heap of crushed rock is simply sprayed with very dilute sulfuric acid and dilute copper sulfate solution trickles from the bottom of the pile.

d) What conditions help this process to go faster?
e) What conditions in this process might have a damaging effect on wildlife and the environment?

Finally, copper metal is extracted from the copper sulfate solution by electrolysis.

Explaining the properties of alloys

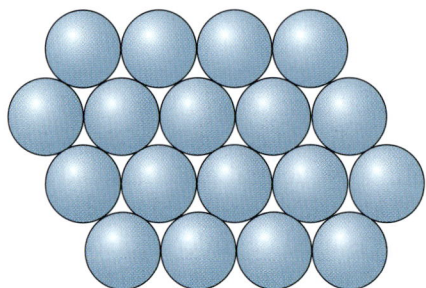

Figure 5.40 Close packing of atoms in a metal

X-ray analysis shows that the atoms in most metals are packed as close together as possible. This arrangement is called **close packing**. Figure 5.40 shows a few close-packed atoms in one layer of a metal.

The bonds between atoms in a metal are strong, but they are not rigid. When a force is applied to a metal, the layers of atoms can slide over each other, allowing the metal to be soft and malleable. This movement of atoms in a metal is called **slip**. After slipping, the atoms settle into position again and the close-packed structure is restored. Figure 5.41 shows the positions of atoms before and after slip. This is what happens when a metal is hammered or pressed into different shapes.

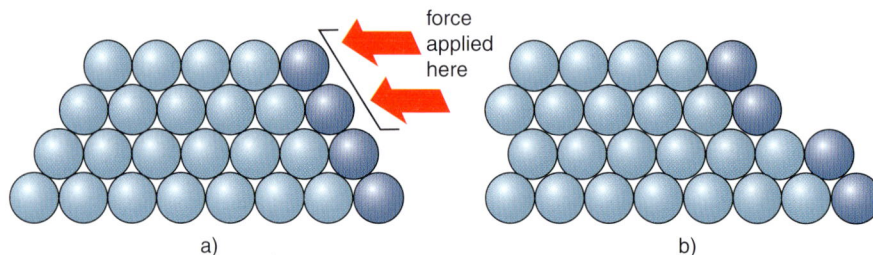

Figure 5.41 The positions of atoms in a metal a) before and b) after 'slip' has occurred

In steel, the regular arrangement of iron atoms is disrupted by adding smaller carbon atoms. The different-sized carbon atoms distort the layers of iron atoms in the structure of the pure metal. This makes it more difficult for the layers to slide over each other, so the steel alloy is harder and much less malleable (Figure 5.42).

force applied here

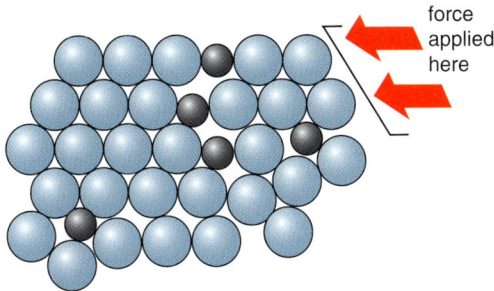

Figure 5.42 Slip cannot occur so easily in the steel alloy because the atoms of different sizes cannot slide over each other

Activity – Recycling metals

Do you and your family recycle metals? Scrap metals such as drinks cans, aluminium foil and old domestic machines can be melted down and used again. Local authorities are now expected to achieve certain targets for recycling waste materials such as metals, glass, paper and plastics. In many areas, there are strong reminders to encourage the recycling of these materials.

1. Extracting metals usually involves four stages (see Figure 5.31). Which of these stages are avoided if metals are recycled?
2. Now look carefully at 'Counting the cost of extracting metals' in Section 5.6. Use this section and Figure 5.29 in the activity 'Quarrying or countryside' to list the advantages of recycling metals.
3. Recycling is not always easy. Scrap metal has to be collected and then transported to where it can be re-used. The metal has to be separated from other material and sometimes one metal has to be separated from another.
   a) Bearing in mind that the metal has to be melted down during the recycling process, how do you think metals are separated from paper?
   b) The two metals recycled in the largest amounts are aluminium and iron (steel).
      How do you think these are separated?
4. Practically all the gold we use is recycled, but only about half the aluminium. Why do you think there is this difference?
5. Find out about the plans and targets for recycling metals in your area and write a few sentences about them.
6. What further initiatives could be taken to improve the recycling of metals in your area?

Further considerations of recycling are covered in Chapter 3.

27. a) Why are the furnaces used to make iron called *blast* furnaces?
    b) Why are blast furnaces usually built near coal fields?
    c) Why is most iron from the blast furnace converted to steel?
    d) Suggest three reasons why steel (iron) is used in greater quantities than any other metal.

28. Explain the following terms: *close packing*; *slip*; *malleable*.

29. The strengths of two metal wires can be compared by measuring the force needed to break each wire using the apparatus in Figure 5.43.

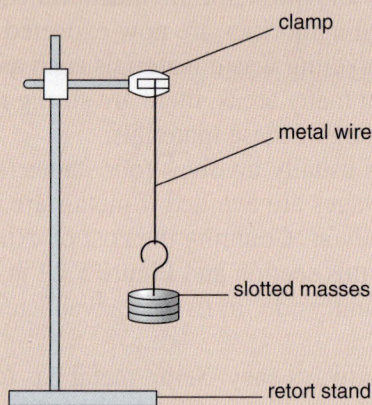

Figure 5.43 Comparing the strengths of metal wires

a) Describe briefly how you would carry out the experiment.
b) What measurements would you record?
c) Copy and complete the following sentences using words from the box below.

> dependent   fair   independent
> measured   same   selected

In this experiment the _____ variable is the type of metal wire used. This is the variable that is changed or _____ by the investigator.
The _____ variable in the experiment is the force needed to break each wire. This is the variable that is _____ when the independent variable changes.
In order to make the experiment _____, it is important that only the independent variable affects the dependent variable. All other possible variables must be kept the _____.

d) State two variables that you would control to ensure the wires are tested fairly.

Summary

✓ **Chemistry** is the study of materials and substances. Chemists and chemical engineers study materials and try to change **raw materials**, like iron ore and limestone, into useful **manufactured materials**, like steel, cement and glass.

✓ • **Elements** are the simplest substances. They cannot be broken down any further.
   • **Compounds** are substances containing two or more elements chemically combined together.
   • **Mixtures** are two or more substances that are mixed together but *not* combined chemically. An **alloy** is a mixture of a metal with one or more other elements.

✓ All substances are made of **atoms**. Each element contains only one sort of atom. Compounds contain two or more different atoms joined together by chemical bonds.

✓ Atoms have a small central **nucleus** around which there are **electrons**.

✓ When atoms combine, they can either:
   • share electrons to form **molecules**,
   • or give and take electrons to form **ions**.

   An **atom** is the smallest particle of an element.
   A **molecule** is a particle containing two or more atoms joined by chemical bonds.
   An **ion** is a charged particle formed from an atom by the loss or gain of one or more electrons.

✓ Elements and atoms of elements are represented by **symbols**. Compounds and molecules of a compound are represented by **formulae**.
   A formula shows the number of atoms of the different elements that are joined together in one molecule of a compound.

✓ In each **group** of the **Periodic Table**, the elements have similar properties.

✓ **Balanced chemical equations** use symbols and formulae to summarise the reactants and products in a reaction.

✓ **Limestone** is mainly calcium carbonate, $CaCO_3$. It is one of the most important and useful naturally occurring building materials. It provides us with quicklime and slaked lime and is a starting point for the manufacture of cement, concrete and glass.

✓ **Extracting metals** from their ores involves four stages:
  ● mining the ore;
  ● concentrating the ore;
  ● reducing the ore to the metal;
  ● purifying the metal.

✓ There are two main methods of extracting metals from their ores:
  ● electrolysis of fused (molten) compounds for metals above carbon in the reactivity series;
  ● reduction of oxides with coke (carbon), or carbon monoxide, for metals below carbon in the reactivity series.

✓ The quarrying and mining of rocks and ores, followed by the extraction of metals, brings both benefits and drawbacks (Table 5.4).

Benefits	Drawbacks
● Useful products and materials ● Creates jobs and employment ● Increases the wealth of a community	● Pollution from smoke, dust and noise ● Destroys wildlife habitats ● Damages the environment with spoil heaps, quarries and mines

Table 5.4 The benefits and drawbacks of quarrying, mining and metal extraction

✓ **Redox** involves reduction and oxidation.
  Oxidation occurs when a substance gains oxygen.
  Reduction occurs when a substance loses oxygen.

✓ The recycling of metals is important because:
  ● it helps to conserve limited resources of the metals;
  ● it reduces damage to the environment;
  ● it saves the cost of extracting the metals.

EXAM QUESTIONS

1 At one time, gutters and drainpipes were made of iron. Today they are made of plastics like PVC (polyvinyl chloride).
  a) What properties of iron made it useful for gutters and drainpipes? *(2 marks)*
  b) What were the problems of using iron? *(2 marks)*
  c) Why has iron been replaced by plastics? *(2 marks)*

2 a) Limestone is used to make some important products. Name two important products from limestone listed below. *(2 marks)*

  cement   diesel   glass   petrol   plastic

  b) In an experiment a student heated a piece of limestone very strongly as shown in Figure 5.44.

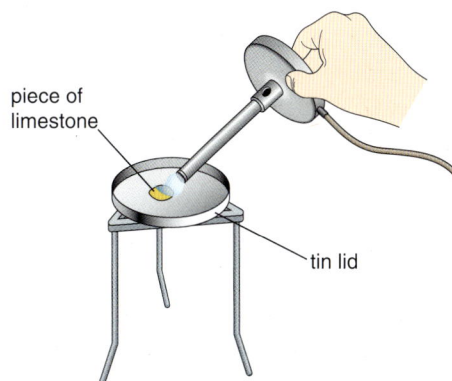

piece of limestone

tin lid

Figure 5.44

  i) State *one* safety precaution that the student should take during this experiment. *(1 mark)*
  ii) When limestone is heated, it forms two products: a white powder and carbon dioxide. What is the chemical name of the white powder? *(1 mark)*

c) The student did a second experiment using 3.00 g of limestone. The limestone was weighed before and after being heated. The student then repeated this experiment using a new sample of 3.00 g of limestone. The results are shown in Table 5.5.

	Experiment 1	Experiment 2
Mass of limestone before heating	3.00 g	3.00 g
Mass of limestone after heating	1.68 g	1.72 g
Mass lost	1.32 g	1.28 g

Table 5.5

i) What is the average mass lost for the two experiments? (*1 mark*)
ii) Why is it important to repeat this experiment? (*1 mark*)
iii) Why is the mass lost not the same for the two experiments? (*1 mark*)
iv) Why is a balance which measures to the nearest 0.1 g *not* suitable for this experiment? (*1 mark*)

❸ Car bodies are made from low-carbon steel.
a) Explain in terms of atoms, why pure iron is useless for making car bodies. (*2 marks*)
b) Steel used for making car bodies is an alloy containing 99.8% iron and 0.2% carbon. Explain in terms of atoms how such a small percentage of carbon makes the steel suitable for car bodies. (*2 marks*)

❹ Look at Figure 5.45.
a) Which process produces the strongest alloy – chill casting (rapid cooling) or sand casting (slow cooling) of the liquid alloy? (*1 mark*)
b) Are chill casting and sand casting examples of discrete variables, ordered variables or categoric variables? (*1 mark*)
c) What percentage of aluminium produces the strongest alloy? (*1 mark*)
d) How many times stronger is this alloy than pure copper? (*1 mark*)

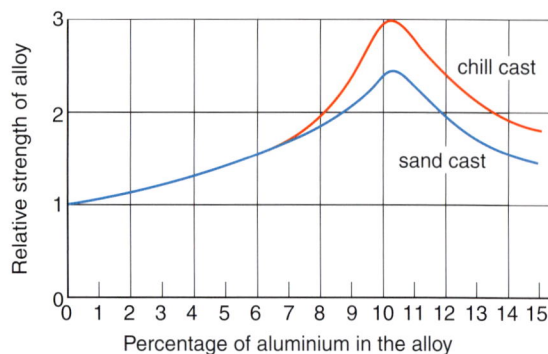

Figure 5.45 The effect of aluminium on the strength of copper alloys

e) What percentage of aluminium produces a sand-cast alloy twice as strong as pure copper? (*1 mark*)
f) Why do you think the strength of the alloy increases at first and then decreases as more aluminium is added? (*4 marks*)

❺ a) A word equation for making a simple glass of calcium silicate ($CaSiO_3$) is:

limestone + sand → calcium + carbon
silicate dioxide

Write a balanced chemical equation for the reaction. (*2 marks*)
b) Glass is used for bottles and ovenware.
i) List four properties which make glass so suitable for these uses. (*4 marks*)
ii) List two drawbacks of using glass for bottles and ovenware. (*2 marks*)

❻ a) Many industrial processes involve the removal of minerals by quarrying. All quarrying has some effect on the environment and on people's lives. Make six important points about the social, economic, health or environmental effects of quarrying. (*6 marks*)
b) Aluminium is more expensive than iron. Why then is aluminium and *not* iron used for the central core of power cables between pylons in the National Grid? (*2 marks*)

Chapter 6
How does crude oil provide useful materials?

At the end of this chapter you should:

✓ understand that crude oil (petroleum) is a mixture of alkanes that can be separated into useful products;

✓ know how the process of oil refining works;

✓ understand that burning fossil fuels such as crude oil causes environmental problems, including acid rain and climate change;

✓ understand how cracking produces different hydrocarbons, which are useful as fuels;

✓ know that cracking produces alkenes, which can be used to manufacture a range of products;

✓ understand that ethene can react with steam to make ethanol;

✓ know that alkenes can be polymerised to make new materials known as polymers;

✓ know that polymers have a range of useful properties;

✓ understand the different uses of polymers;

✓ know that some polymers can be recycled easily;

✓ understand that most polymers are not biodegradable;

✓ be able to evaluate the effects on the environment of burning hydrocarbon fuels;

✓ know about the development of better fuels;

✓ be able to evaluate the uses of crude oil as a fuel and as a chemical to manufacture other materials;

✓ know about the development of polymer materials;

✓ be able to evaluate the manufacture of ethanol from renewable and non-renewable sources.

Figure 6.1 In the last 50 years, the 'garage' has changed its function. Gone are the mechanics and the repair of cars. The garage is now where you get your milk, lottery ticket, newspaper, sweets, phone top-up card, and fuel for the family car. Garages are an essential part of our everyday living and not just for the 'petrol heads'. The oil industry touches our lives many times a day. Often there are benefits, sometimes there are drawbacks.

Crude oil

Crude oil is a naturally occurring material found underground. It is a mixture of very similar compounds, called **hydrocarbons**.

Crude oil forms over thousands, if not millions, of years as dead organisms slowly break down. The hydrocarbons in crude oil vary a great deal in molecular size. These different sized molecules can be separated by fractional distillation.

Distillation

Mixtures of liquids can be separated by **distillation**.

Different substances have different boiling points. When a liquid boils, the liquid particles have to be separated from each other to make a gas. Three factors affect the boiling point of a liquid:
- atmospheric pressure – substances boil at a lower temperature under reduced pressure;
- the mass of the molecules in the liquid;
- the strength of the forces between the liquid particles.

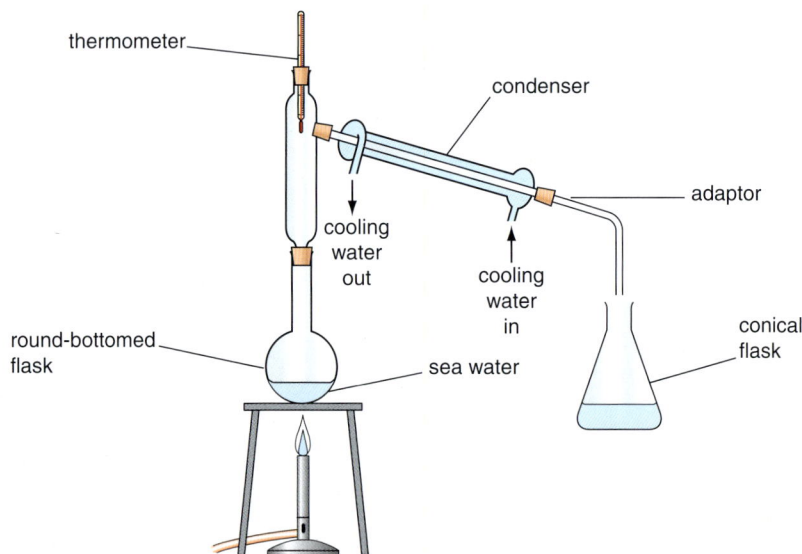

Figure 6.3 A liquid can be separated from a solution using this apparatus. Water or air can be used to cool the condenser. In this case, the condenser is cooled by water

When a simple mixture like salt water (sodium chloride solution) is boiled, only water molecules are turned into vapour. The sodium ions and chloride ions in the water do not vaporise. They remain in the boiling liquid. The vapour that boils off is pure water. This can be cooled and condensed back to a liquid.

Various distillation methods are used to produce pure water from salt water. One of these is shown in Figure 6.4.

> A **hydrocarbon** is a compound that contains hydrogen and carbon atoms only.

Activity – Mixtures

❶ Crude oil is sticky stuff. It often gets mixed with other materials. How could you separate the following mixtures:
  a) crude oil and seawater, to get clean seawater;
  b) crude oil and seawater, to recover the crude oil;
  c) crude oil and sand, to get clean sand (you may need to use another chemical)?

❷ a) How would you clean crude oil off sea-bird feathers?
  b) What adverse effects might this have on the bird?

Figure 6.2 The bird may not appreciate this treatment!

Fractional distillation

When a liquid boils, or vaporises, its particles gain enough energy to escape from the surface and become a gas (called a vapour). **Distillation** occurs when this vapour is removed and condensed back to liquid in a separate container.

Ordinary distillation cannot separate liquids with boiling points that are close together. This is done by **fractional distillation**.

For example, when a mixture of ethanol (boiling point 78 °C) and water (boiling point 100 °C) is heated, both the ethanol and the water molecules will **vaporise** together. When the vapour is condensed, it is still a mixture but there is more ethanol than water in this condensed mixture because ethanol boils at a lower temperature. The ethanol molecules 'escape' from the boiling liquid more easily.

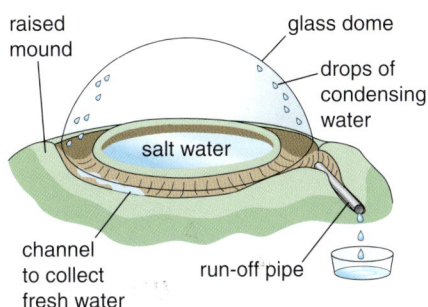

Figure 6.4 Turning salt water into pure water – sunshine is often used as an energy source

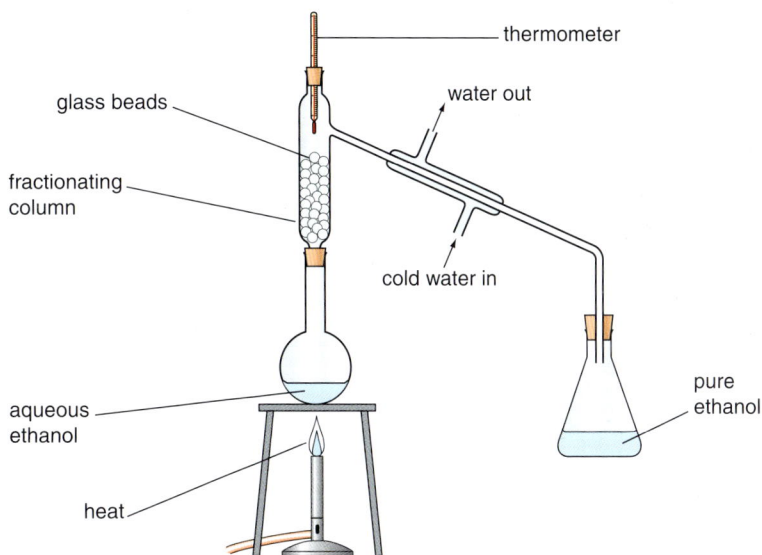

Figure 6.5 During fractional distillation, the vapours condense and evaporate several times in the vertical column

During fractional distillation, hot vapour condenses on cooler surfaces inside the column. The condensed mixture then evaporates again as more hot vapour rises up the column. Each time this happens, more of the lower boiling point liquid evaporates, and the higher boiling point liquid is left behind.

The higher boiling point liquid runs back down the column, and into the boiling mixture.

After the vapour has evaporated and condensed several times on its way up, only the lower boiling point liquid reaches the top of the column. This flows out of the column and is collected.

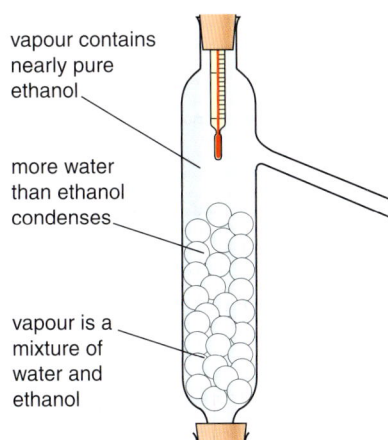

Figure 6.6 How a fractionating column works

Alkanes

The simplest hydrocarbons are alkanes. When carbon atoms form compounds, they can form four chemical bonds with other atoms, but hydrogen atoms can form only one chemical bond. So, the simplest alkane of all is methane, CH_4 (Figure 6.7).

The hydrocarbon with two carbon atoms is shown in Figure 6.8. It is called ethane.

Can you see that ethane is like methane with an extra —CH_2— unit? If you add another —CH_2— unit you get a heavier molecule called propane (Figure 6.9a). Another —CH_2— unit makes butane (Figure 6.9b).

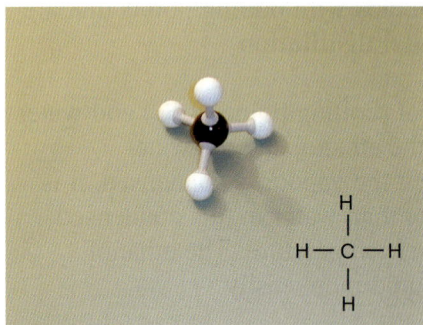

Figure 6.7 The structure of methane

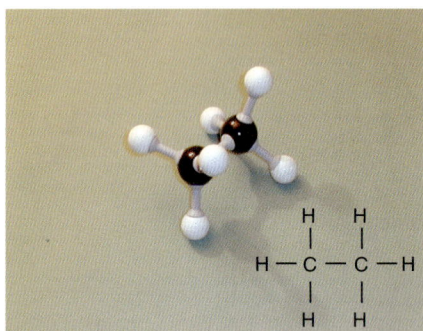

Figure 6.8 The structure of ethane

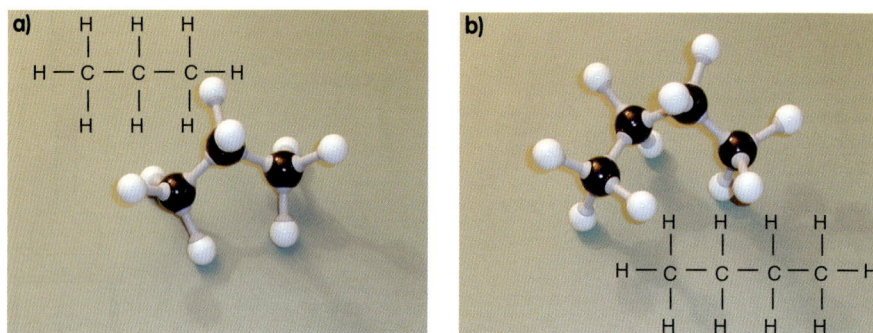

Figure 6.9 The structure of a) propane and b) butane

It is easy to see that this is becoming a series of similar molecules. This series of hydrocarbons are called the **alkanes**. Any alkane will have a chemical formula that can be written as C_nH_{2n+2}, where 'n' is a whole number. C_nH_{2n+2} is known as the general formula for alkanes. All the alkanes have similar properties because of their similar molecular formulae.

The most important property of alkanes is that they burn in air to form carbon dioxide and water vapour. This is why the alkanes are all used as fuels (Table 6.1).

Alkane	Use
Methane	Kitchen gas for cookers
Propane	Camping (calor) gas for cookers
Butane	Bottled gas for heating
Octane	Petrol for cars
Dodecane	Diesel for vans

Table 6.1 Alkanes as fuels

Activity – Is there a link between the size of the molecules in an alkane and its boiling point?

Alkane	Formula	B Pt / °C
Methane	CH_4	−161
Ethane	C_2H_6	−89
Propane	C_3H_8	−42
Butane	C_4H_{10}	0
Pentane	C_5H_{12}	36
Hexane	C_6H_{14}	69
Heptane		98
Octane	C_8H_{18}	125
Nonane		151
Decane	$C_{10}H_{22}$	174

Table 6.2 Boiling points of the alkanes

❶ Write down the missing formulae of heptane and nonane.
❷ Plot a graph of the boiling points for all the alkanes in the table against the number of carbon atoms in the molecules.
❸ a) What pattern does the graph show?
   b) Why is there a pattern?
❹ Estimate the boiling point of the alkane $C_{12}H_{26}$.
❺ The boiling point of a pure substance is a very 'reliable' piece of data. This means that if you repeat the measurement you get the same result time after time. Is the pattern in the graph reliable? Explain your answer.

Activity – How much energy is transferred when 1 g of hydrocarbon is burned?

Plan an investigation to find out how much energy is transferred when 1 g of a hydrocarbon is completely burned in air (oxygen). You could use:
- camping gas as a source of butane, C_4H_{10};
- paraffin oil as an alkane with about 12 carbon atoms in each molecule;
- engine oil as a long chain hydrocarbon.

If you have sufficient time, you can also compare how easily and how cleanly these alkanes burn.

You should always use chemicals from an education supplier, rather than commercial products, for this type of investigation. Rooms should be well ventilated for these experiments and you must always wear eye protection.

Saturated and unsaturated hydrocarbons

Ethane (C_2H_6) and propane (C_3H_8) (Figure 6.10, page 108) are saturated hydrocarbons.

In alkanes, all the carbon atoms are joined together with single carbon–carbon bonds. These bonds (see Section 5.2) are made as the carbon atoms share their electrons with each of the carbon atoms next to them. This results in a very stable skeleton.

Alkanes, in which all the carbon atoms are joined by single bonds, are called **saturated hydrocarbons**.

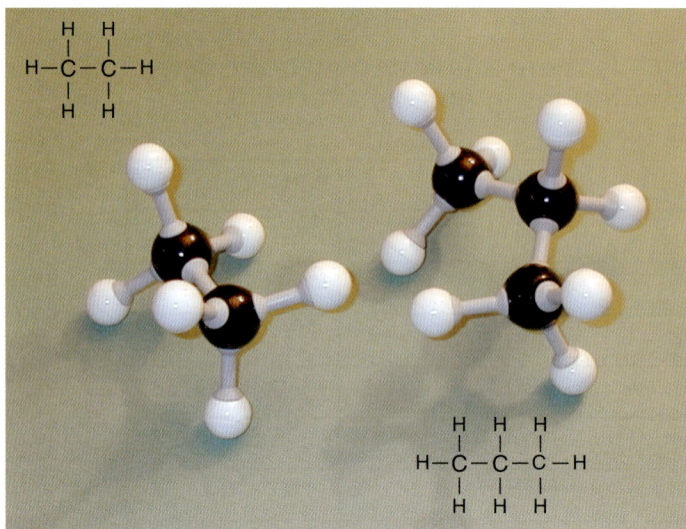

Figure 6.10 Structures of ethane and propane. Remember the plastic model is only a representation of the molecule. It is not a 'real' picture of the structure

Saturated hydrocarbons burn in air, but they don't react with many other chemicals.

Ethene (C_2H_4) and propene (C_3H_6) (Figure 6.11) are very similar to ethane and propane, except that they contain a **double bond** between two of the carbon atoms in their molecules. These hydrocarbons containing a double carbon–carbon bond are called **unsaturated hydrocarbons**. Unsaturated hydrocarbons with a carbon–carbon double bond are also called **alkenes**.

❶ How many chemical bonds can a carbon atom make with other atoms in a molecule?

❷ Predict the chemical formulae of the alkenes with 15 carbon atoms and 20 carbon atoms.

❸ What is meant by a saturated hydrocarbon?

❹ What is meant by an unsaturated hydrocarbon?

❺ 'Black and white molecular models are useful for thinking about chemical reactions even though they are models and not real molecules.' Explain whether you agree or disagree with this statement.

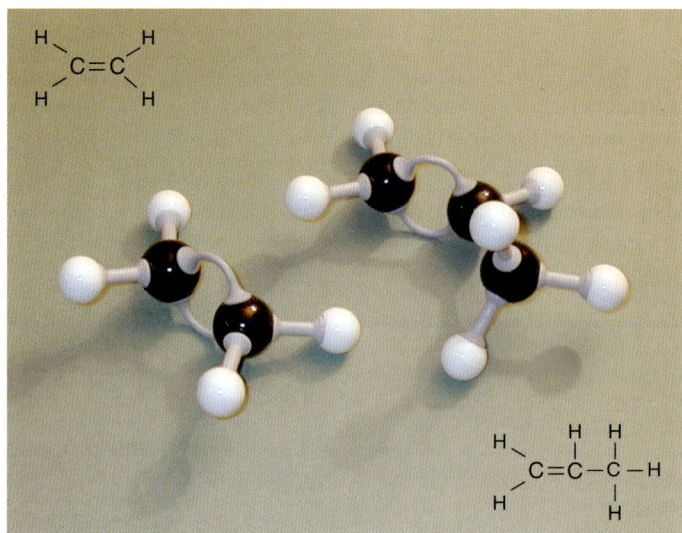

Figure 6.11 Ethene and propene – unsaturated hydrocarbons

The carbon–carbon double bond is a very reactive area in these molecules. It can take part in lots of reactions. Unsaturated hydrocarbons are therefore much more reactive than saturated hydrocarbons.

6.3 Oil refinery

An oil refinery is a large chemical works which processes crude oil and separates it into useful products. As it comes out of the ground, crude oil is a useless mixture of hydrocarbons.

The evaporation and condensation of substances inside the fractionating tower sorts molecules of the crude oil mixture into separate groups. The biggest molecules remain on the lower levels of the tower where the temperature is highest. The smaller molecules don't condense so easily. They travel up the tower to higher levels where they condense at lower temperatures. The products that are taken from the fractionating tower at different levels are called **fractions**. The different 'fractions' are *not* pure substances. They are just mixtures of substances with similar boiling points and similar-sized molecules.

Figure 6.12 The fractionating tower in an oil refinery works like a fractionating column in the laboratory, but on a much larger, industrial scale

A **fraction** from an oil refinery (for example, naptha, kerosene) contains a mixture of compounds with molecules of similar size and boiling point.

6 How is the petroleum gas fraction used *in the refinery*? See Table 6.3 on page 110.

7 What is the naphtha fraction used for?

8 Why can you not get pure octane from the fractionating tower?

9 Look carefully at Table 6.3.
   a) What is the trend in the colour and viscosity of the fractions as you move down the fractionating tower?
   b) Explain the trend in viscosity.
   c) How does the trend in colour and viscosity relate to the size of the molecules in the fractions?

10 How does the trend in flammability (how easily it burns) relate to the size of the hydrocarbon molecule in a fraction?

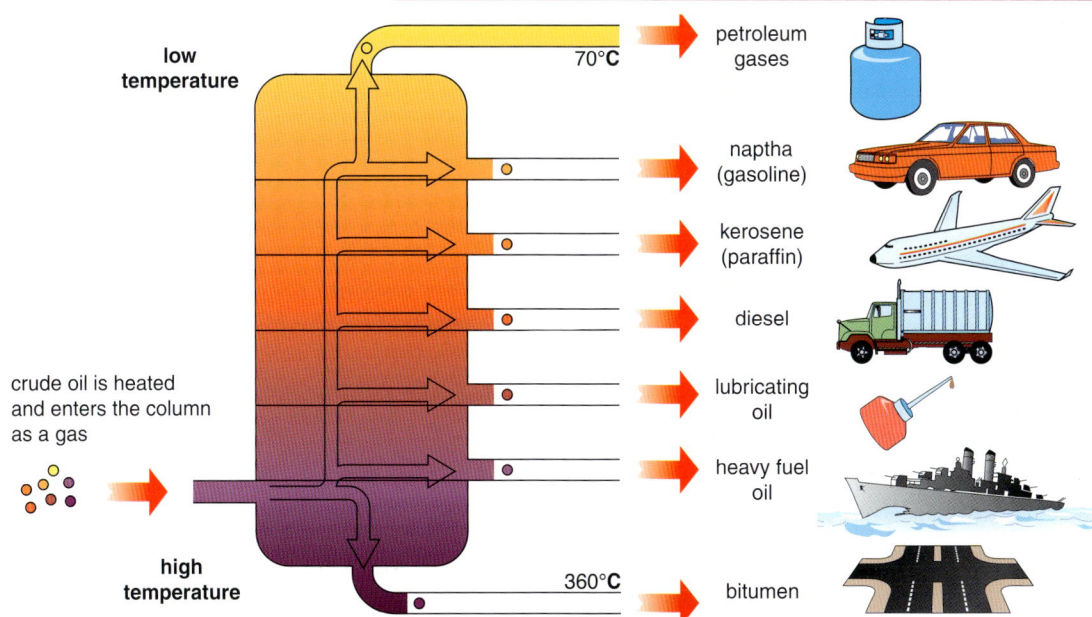

low temperature

70°C — petroleum gases

naptha (gasoline)

kerosene (paraffin)

diesel

crude oil is heated and enters the column as a gas

lubricating oil

heavy fuel oil

high temperature

360°C — bitumen

Figure 6.13 A fractionating tower in an oil refinery

Fraction	Number of carbon atoms in the molecule	Description	B Pt (°C)	Flammability	Uses
Natural gas	1–4	Colourless gas	Less than 40	Explodes if mixed with air and lit	Used as a fuel in the refinery. Bottled and sold as LPG
Naphtha	5–10	Yellowish liquid flows very easily	25–175	Evaporates easily, vapour mixed with air is explosive	Petrol. Also used for making other chemicals
Kerosene	10–14	Yellowish liquid flows like water	150–260	Will burn when heated	Aircraft fuel
Light gas oil	14–20	Yellow liquid thicker than water	235–360	Needs soaking onto a wick or other material to burn	Diesel fuel
Heavy gas oil	20–50	Yellow brown liquid	330–380	Only burns when soaked onto a wick – very smoky	Used in the catalytic cracker (see Section 6.5)
Lubricants (motor car engine oils)	50–60	Thick brown liquid like syrup	340–575	Needs to be hot and soaked onto a wick before it burns	Grease for lubrication. Used in the catalytic cracker
Fuel oil	60–80	Thick brown sticky liquid	above 490	Needs to be hot and soaked onto a wick before it will burn	Fuel oil for power stations and ships
Bitumen	more than 80	Black semi solid	above 580	Hardly burns at all, unless very hot	Road and roof surfaces

Table 6.3 Fractions from an oil refinery

Activity – Rocville

Rocville is an imaginary town in a rural area on the west coast of Britain. At one time, Rocville attracted tourists, but cheap flights and cheap accommodation in more exotic destinations has led to reduced numbers of visitors.

A big oil company wants to purchase a deep-water harbour just to the north of Rocville in order to build a new oil refinery terminal there.

Different people from the surrounding area have different views about the proposal. Here are some of their views.

A Rocville town councillor
'The town cannot provide employment for all our young people so they are leaving to find work elsewhere. We need to bring industry to the area, and an oil refinery would be a good option. It will make our town more prosperous and attract other businesses.'

A Rocville fishing boat owner
'The deep-water estuary is one of our best fishing grounds. My family have been fishing these waters for two hundred years. The big tankers will pollute the water and kill the fish. Using seawater as a coolant in the refinery will destroy the balance of life in the sea.'

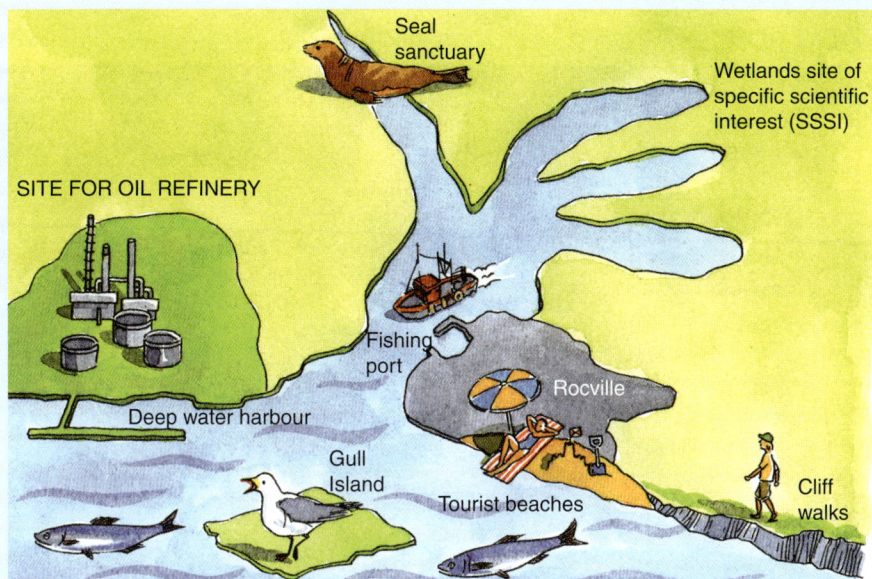

Figure 6.14 Rocville and the surrounding area

The environmental warden for the coast
'Crude oil is a major problem if it contaminates coastal waters. There is a seal sanctuary on our coastline and rare birds nest along the cliffs. The offshore oyster beds and salt marshes that support rare species will be destroyed.'

An executive from the oil company
'This site is ideal for us. The risk of environmental damage from a spill is relatively low. It is true that there is a decline in local air quality near a refinery, but we will bring people and employment to the area.'

An unemployed 17 year old from Rocville
'My life is rubbish. The holiday business only has jobs for the summer months. All I want is a regular job and money. I can put up with the smells of an oil refinery if I get that.'

A young mother with a 2-month-old baby
'I've read stories in the paper that lots of the chemicals in crude oil can cause cancer. I visited a refinery with my school. It smelt really foul.'

A retired person living in Rocville
'We came here after I retired from the saw mill. My wife was a classroom assistant. We want a quiet life and healthy sea air. I have bad asthma after working around sawdust and could not stay here with the fumes from the refinery – it would be the death of me.'

Think about the views of other people in the area who would be affected by a new oil refinery.

❶ Write a short statement that might be made by a young chef working in a Rocville restaurant. He doesn't want to see the Rocville holiday trade vanish, but is keen for the town to prosper.

❷ A local landowner has phoned the Rocville radio station about the proposal. Write what she might say on air. She'd like to make money from selling some land, but she's also worried about her farm workers leaving for higher wages elsewhere.

❸ Work in groups of three or four for this question. You are the council planning officers for Rocville. Write a briefing paper to advise the local council on the proposal. In your briefing paper:
   a) collect together all the arguments in favour of the refinery;
   b) collect together all the arguments against the refinery;
   c) add a short introduction at the beginning of your report, giving some background information;
   d) make a decision about the refinery as a group and write this as your recommendation at the end of the paper. If you have access to ICT, add pictures and other graphics. You could make the briefing paper into a presentation and present it to your class.

Air pollution
Acid rain

Figure 6.15 Over time, acid rain pollution can destroy beautiful monuments

The worst acid rain ever recorded was in 1983 at the Inverpolly Forest, Scotland. The pH value recorded for the acid rain was 1.87. The rain was more acid than strong vinegar and would turn Universal Indicator red. Unpolluted rainwater is naturally acid and always has been. Rainwater reacts with carbon dioxide in the air to form a weak acid. The acid in unpolluted rain is carbonic acid.

During the last 200 years, the problem of acid rain has got much worse. Pollutants from industrial and domestic processes have made rainwater more acidic. Acid rain attacks marble and decorative stonework. Worse still it kills vegetation and runs off into rivers and lakes where it kills the water life.

When fossil fuels are burnt, the sulfur atoms in the fuels eventually end up as sulfuric acid. When petrol and diesel burn at high pressure in vehicle engines, oxides of nitrogen are formed. These react with rainwater to make nitric acid. Sulfuric acid and nitric acid are much stronger acids than carbonic acid. They give the acid rain a much lower pH.

Climate change and global warming

The Earth's surface is getting warmer. Scientists predict increasing extreme weather and say there is growing evidence that human activity, particularly burning fuels, is to blame.

What is the 'greenhouse effect'?

The 'greenhouse effect' is caused by a layer of gases which trap heat from the Sun in the Earth's atmosphere. Without them, the planet would be too cold to sustain life as we know it. These gases include carbon dioxide which is released by the burning of fossil fuels.

---

⑪ What is Universal Indicator solution used for?

⑫ What colour is Universal Indicator solution in a) pure water; b) strong acid?

⑬ What acid is naturally present in unpolluted rain?

⑭ What acids are present in acid rain?

⑮ Write word equations for the formation of the acids which pollute rainwater.

⑯ Catalytic converters in car exhausts can convert oxides of nitrogen into nitrogen gas. Explain why this reduces acid rain.

⑰ Explain how sulfur dioxide scrubbers in factories and low sulfur diesel fuels for lorries help to solve the problem of acid rain.

⑱ Design a poster to explain the effects of acid rain to a non-reader.

What is the evidence for global warming?

Temperature records show that average global temperature increased by about 0.6 °C in the twentieth century. Sea levels have risen by 10–20 cm and Arctic sea-ice has thinned by 40% in recent decades.

If nothing is done to reduce 'greenhouse gas' emissions, scientists predict a global temperature increase of about 5 °C by 2100. There will be more rainfall overall but droughts in inland areas will increase. More flooding is expected from storms and rising sea levels. Poorer countries, which are least equipped to deal with rapid change, will suffer most.

Other factors may slow down the warming – such as **global dimming** or plants taking up more carbon dioxide as their growth rate is increased by warmer conditions. Scientists are not sure how the complex balance between the positive and negative effects will play out.

What about the sceptics?

Most global warming sceptics do not deny that the world is getting warmer. But they do doubt that human activity is the cause. Some say the changes, now being witnessed, are not extraordinary because similar, rapid changes have occurred at other times in the Earth's history. There is more on global warming in Chapter 3.

Global dimming happens when particles in the atmosphere prevent sunlight reaching the Earth. Soot and ash particles from burning fossil fuels, ocean spray and aircraft contrails all add to particles in the air.

19 Why are tiny particles in the air particularly harmful to those with breathing problems?

20 a) What evidence is there that global warming is just a natural process?
b) What evidence is there that contradicts this view?

21 Which of the following three predictions do you think will actually happen? Write a paragraph to explain your answer.
a) We'll worry and blame ourselves for climate change for thousands of years.
b) Fossil fuels will run out and renewable energy will save us.
c) The oceans will evaporate as the Earth heats up and humans will all die.

Activity – Acid rain

This table shows the emissions of sulfur dioxide from large combustion plants and the total sulfur dioxide emissions in thousands of tonnes per year.

Year	1970	1980	1990	1995	2000	2001	2002	2003
Emissions from large combustion plants	3717	3449	2935	1756	899	822	745	734
Total emissions	6456	4841	3711	2354	1194	1118	1002	979

1 a) Plot a graph of the emissions from large combustion plants, along the y-axis, against the year, along the x-axis. Take care with the x-axis scale; make sure you space out the years correctly. Plot the y-axis scale from 0 to 4000. Remember that this scale is thousands of tonnes of sulfur dioxide per year.
b) Draw an appropriate line through the points on the graph.

2 An industry spokesman commented on this data. He said:
'The acid rain problem of the 1970s and 80s is over. There is no longer anything to worry about. The industry has put its house in order. The sulfur dioxide emissions from large combustion plants are a thing of the past. Our acid gas emissions from burning fossil fuels are now insignificant and not a pollution hazard.'
a) Do you agree with the industry spokesman?
b) Write a paragraph to explain why you agree or disagree with what he said.

Major sources	1970	1980	1990	2000	2001	2002	2003
Road transport	42	51	61	40	39	39	38
Residential heating	209	93	47	27	29	27	20
Energy and other industries	86	83	74	25	21	12	13
Total	486	332	287	168	169	150	141

Table 6.4 Emission of PM$_{10}$s in thousands of tonnes per year

Particles that are less than 10 micrometres in size (PM$_{10}$s) are produced when fossil fuels burn. The particles are too small to see. They are also too small to settle to the ground and remain suspended in the air.

Table 6.4 gives some data about PM$_{10}$ emissions.

1. Display the information about total emissions in a suitable way so that it can be used in a presentation.
2. Draw charts to compare the PM$_{10}$ emissions from the different sources in 1970, 1990 and 2003.
3. The decline in residential heating emissions over the period of 1970–2003 can be explained by the reduction in the use of coal for domestic heating. What forms of heating have replaced coal, and why has this led to a reduction in emissions?
4. Look at the trends in the data. In your opinion, should transport and energy industries be congratulated on the progress they have made to reduce this source of air pollution?

6.5 Catalytic cracking

Light fractions	Use
Gas Naphtha / Petrol Kerosene Diesel fuel	All can be used directly as fuels without any processing
Heavy fractions	
Light & heavy gas oil Lubricants Fuel oil Bitumen	These cannot be used as fuels except in large engines such as those on a ship, where the oil is heated up before burning

Table 6.5 Comparison of the flammability of light and heavy fractions from crude oil

Cracking is the name given to the breaking up of large hydrocarbon molecules into smaller and more useful bits. You can do this in the laboratory at school using high temperatures and a catalyst, but in the oil industry they also use high pressures because it is more efficient.

Figure 6.16 Heavier fractions from crude oil are not much use as a fuel

The process is controlled by careful choice of conditions to produce mixtures of smaller hydrocarbons. Some of these smaller hydrocarbons are alkenes with carbon–carbon double bonds.

For example, during cracking, the hydrocarbons are vaporised and passed over the catalyst at 500 °C. The temperature and pressure used for the cracking process are selected to give a high percentage of hydrocarbons with 5–10 carbon atoms. These are the hydrocarbons needed for petrol.

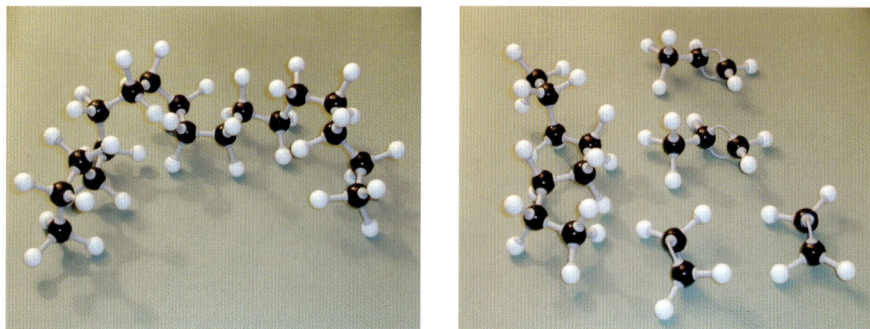

Figure 6.17 When a large molecule like $C_{18}H_{38}$ is cracked, you can end up with several fragments. Octane (C_8H_{18}) is one of the molecules found in petrol. Ethene (C_2H_4) and propene (C_3H_6) are useful starting substances for the manufacture of plastics and other chemicals

Figure 6.18 Which of these pasta shapes can become the most tangled?

Viscosity measures how easily a liquid can flow. Thick liquids have a high viscosity. Runny liquids have a low viscosity.

22 What is meant by a catalyst?

23 Why are the heavier fractions from crude oil unsuitable for use as fuels?

24 'Cracking produces more useful molecules.' Explain this statement.

25 Draw a flow diagram to summarise the catalytic cracking process.

26 Catalytic cracking is a thermal decomposition reaction. Why is heating needed to start the reaction?

27 Write a balanced equation for the complete combustion of octane (C_8H_{18}).

28 Copy and complete this equation for the *incomplete* combustion of $C_{11}H_{24}$ in a limited supply of air.

$$C_{11}H_{24} + 6O_2 \rightarrow$$

Viscosity

Some hydrocarbons have long, flexible molecules. The longer the molecules, the more tangled they become with other molecules (Figure 6.18). This makes the liquid thicker (more viscous) and less easy to pour. Motor oils have a range of **viscosities** displayed on their containers. They are written as numbers with a 'W' after them.

Alkenes

In Section 6.2, we noticed that the alkanes formed a series of similar hydrocarbons. The alkenes also form a series of similar compounds. Each alkene has one carbon–carbon double bond in its molecule. So, the simplest alkene is ethene (C_2H_4). If you add one —CH_2— link to this molecule, it becomes propene (C_3H_6), the next molecule in the series. Add more —CH_2— links and the next three members of the series are butene (C_4H_8), pentene (C_5H_{10}), and hexene (C_6H_{12}). This makes a series of similar compounds, with a general formula of C_nH_{2n} for the alkenes.

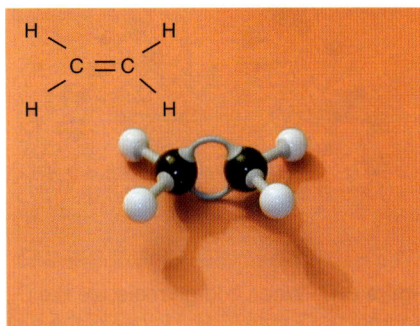

Figure 6.19 The molecular model of ethene

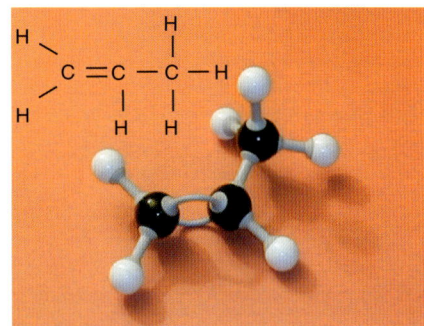

Figure 6.20 The molecular model of propene

The reactivity of alkenes

The carbon–carbon double bond in alkenes is a target for reactions. One of the two bonds can easily be broken, allowing the carbon atoms to link with other atoms. For example, propene will act as in Figure 6.21 to 6.23 in the presence of a reactive substance such as bromine (Br_2).

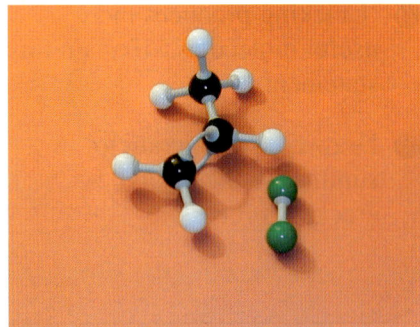

Figure 6.21 The green molecule is a molecule of bromine that can react with propene.

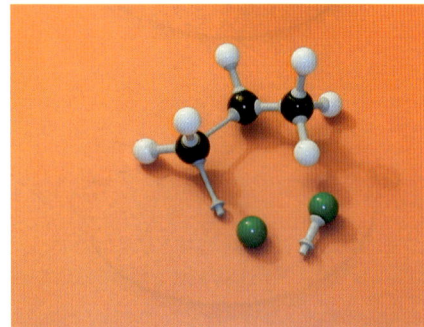

Figure 6.22 When propene and bromine water react, one of the bonds in the double bond of the propene molecule opens up. Each bromine atom then bonds with a carbon atom.

When one of the bonds in the double bond of the propene molecule has opened up, the two atoms in the bromine molecule are added to the propene molecule.

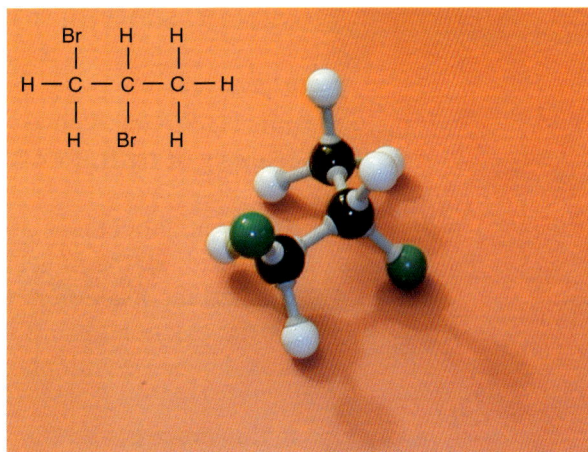

Figure 6.23 The product of the reaction has two bromine atoms added to the alkene. The product of the reaction is a saturated compound because it contains only carbon–carbon single bonds

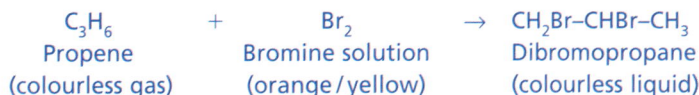

C_3H_6	+	Br_2	→	CH_2Br–$CHBr$–CH_3
Propene		Bromine solution		Dibromopropane
(colourless gas)		(orange / yellow)		(colourless liquid)

㉙ Draw one possible structure for decene ($C_{10}H_{20}$).

㉚ Why is a carbon–carbon double bond more reactive than a single bond?

6.7 Spirit fuel

NEWS: Brazil has started to export 'alcohol motor fuel' based on ethanol spirit made from sugar cane

Brazilian energy giant Petrobras made its first shipment of fuel alcohol this week. The first shipment of alcohol left on Monday from Rio de Janeiro and was destined for Venezuela. A monthly shipment of around 25 000 cubic metres of the fuel is planned.

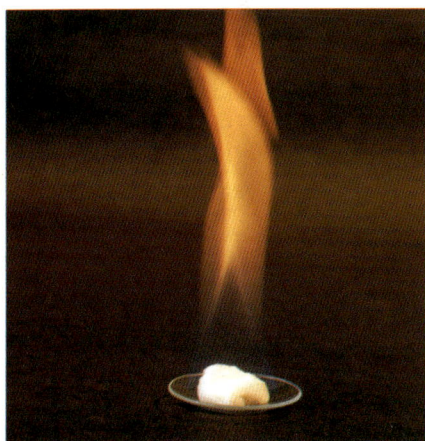

Figure 6.24 Here, some cotton wool has been soaked in ethanol and set alight. Ethanol is clean, safe and carbon neutral

Ethanol has a lot of benefits as a fuel.
- It burns with a clean, smokeless, hot flame.
- It is a liquid so it can be stored and transported easily.
- It is not oily and can be washed away with water.
- It mixes with water and does not pollute it.
- It is present naturally in the environment, for example, in rotting (fermenting) fruit.
- It is 'carbon neutral' when produced from plant products. The same amount of carbon dioxide is released into the air when it burns as that which came out of the air when the plant crop was growing.
- Cars do not need modification to run on it.

There are two methods of producing ethanol: either from ethene or from plants. Let's compare the two methods.

Making ethanol from ethene

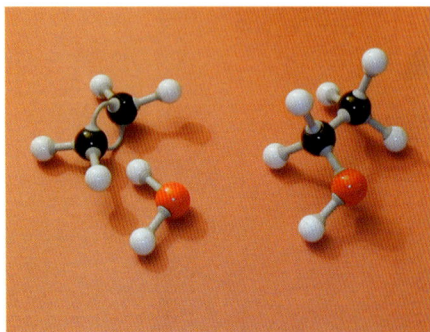

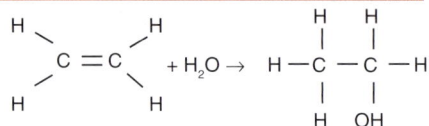

It is possible to make ethanol by adding the atoms from a water molecule to the double bond in an ethene molecule. This reaction is an example of **hydration** because it involves adding water.

Steam and hot ethene are mixed under pressure at 300 °C in a reactor. Even with all the molecules squashed together and colliding with one another, the reaction is very slow. We can reduce the energy required for the reaction to start using a catalyst. The catalyst used is solid silicon dioxide coated with phosphoric acid.

In this process

- Two volumes of ethanol vapour are mixed with one volume of steam at 300 °C.
- This is compressed to 65–70 times atmospheric pressure, then it is passed over the catalyst in the reaction chamber.
- The gases are cooled and ethanol and water condense out.
- Much of the ethene is unchanged – this is recycled and returns to the start of the process.
- The ethanol and water mixture is separated by fractional distillation.
- The water is recycled through the process.

In spite of the extreme conditions in the reactor, only 5% of the ethene is converted into ethanol. By removing ethanol from the mixture and recycling the ethene, it is possible to achieve an overall conversion of 95%.

Figure 6.25 The hydration of ethene. The reaction adds a water molecule to an ethene molecule to make ethanol

Making alcohol by fermentation of plant products

Starchy material such as barley is used as the starting material for this process. Sugar cane is used in hot countries. Starch is a complex carbohydrate. The starch must first be broken down into sugars. The starchy material is heated with hot water to extract the starch and then turned into sugars using natural enzymes in the plant material.

Yeast is then added and the mixture is kept warm at 35 °C for several days until fermentation is complete. Enzymes in the yeast convert the sugars into ethanol and carbon dioxide. This is the fermentation process. The yeast is killed when the ethanol concentration reaches 15%.

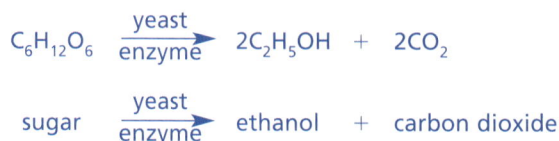

$$C_6H_{12}O_6 \xrightarrow[\text{enzyme}]{\text{yeast}} 2C_2H_5OH + 2CO_2$$

$$\text{sugar} \xrightarrow[\text{enzyme}]{\text{yeast}} \text{ethanol} + \text{carbon dioxide}$$

Finally, ethanol is separated from the watery mixture by fractional distillation.

③① Make a list of the benefits of ethanol fuel over petrol. Put the most important benefits at the top.

③② Explain the term 'carbon neutral'.

③③ Comment on the following statement: 'The source of ethene, to make ethanol, is crude oil, so ethanol made from ethene is just another fossil fuel.'

③④ Ethene does not react easily with water.
a) Why is the reaction carried out at a high temperature?
b) Why is the reaction carried out at high pressure?

c) Why is a catalyst needed?

③⑤ Why must the ethene be passed through the reactor again and again?

③⑥ What is the starting material for making ethanol by fermentation:
a) in a country like the UK;
b) in a hot tropical country?

③⑦ Draw a flow chart to summarise the production of ethanol by fermentation. Start the flow chart with barley grain in the fields and end with ethanol.

Activity – Spirit fuel factory

Ethanol has a part to play as a motor fuel in our future, but at present not enough ethanol is produced to satisfy even half our needs.

❶ Compare the two methods described in Section 6.7 for producing spirit fuel (ethanol). List the benefits and drawbacks of each method. Write

an evaluation of each method and decide which is better.

❷ Devise a plan for setting up a small factory to produce ethanol fuel for the petrol stations in a small town.
Write down a) the starting materials you would need; b) the processes you would carry out; c) how you would store the ethanol fuel before delivery to the petrol stations.

6.8 # Polymers

Figure 6.26 High-quality music on plastic!

Our music collections are important to all of us. 'Records' are what they used to be called – that now sounds really old-fashioned. About 20 years

Plastics are manufactured materials consisting of molecules with long chains of carbon atoms.

A **monomer** is a small molecule capable of forming bonds, or links, with other small molecules to form a larger molecule called a polymer.

A **polymer** is a large molecule made from lots of monomers joined together. Polymers can be made from thousands of monomer units. They can be natural materials such as cellulose or manufactured materials such as polystyrene.

ago, records started to be replaced by CDs. Records and CDs are made of 'vinyl'. Vinyl is a manufactured **plastic**. Without plastics, we wouldn't be able to enjoy recorded music.

Plastics

Plastics are materials called **polymers**.

Polythene

Polythene is the material used for supermarket carrier bags and for most types of flexible packaging. The problem with polythene is that it is not biodegradable. This means that it is not broken down by bacteria, like paper and wood.

Figure 6.27 Bags, and other items made of polythene, litter our streets. They get into rivers, and the sea, after being thrown away. They can remain there for years and years

Polythene is made from ethene. Under high pressure, ethene molecules are so squashed together that they react and take up less space. The double bonds open up and link the molecules together.

Thousands of molecules can be linked in this way creating a very long molecule of polythene.

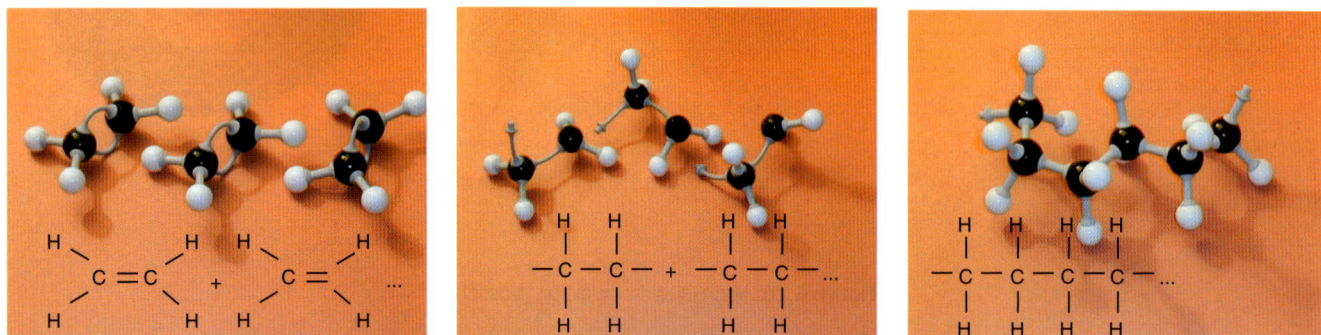

Figure 6.28 We can use plastic models to help us understand how polythene molecules are formed from ethene

Choosing the monomer to give the best plastic

Polythene (or polyethene) is made of long, thin and very flexible but strong molecules.

Polypropene (polypropylene) is a more rigid plastic that keeps its shape better than polythene. It is used for washing up bowls and plastic toolboxes. Although it is stiff, thin pieces are still flexible, so it can be used to make its own hinge. Other **monomers** give different properties to the polymers they form.

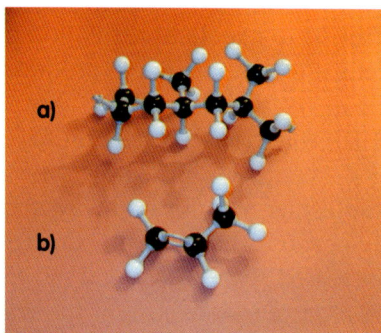

Figure 6.29 Molecular models of propene and part of polypropene

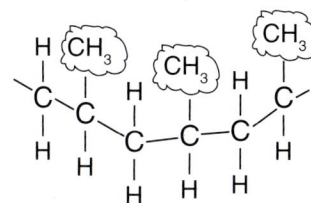

Figure 6.30 Polypropene has better rigidity than polyethene. The CH_3 groups make it less flexible.

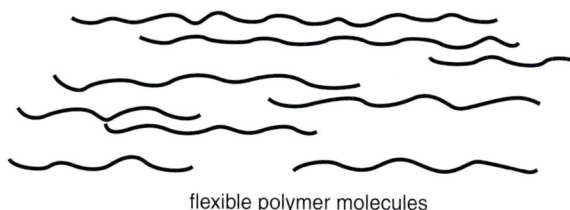

flexible polymer molecules

crosslinks make the structure more rigid

Figure 6.31 Polythene and polypropene are made of long flexible molecules. In rigid polymers, chemical bonds have formed between neighbouring molecules. These bonds join different parts of the polymer together in a three dimensional network. The crosslinks lock the molecules together and stop them from being flexible

38 Explain the difference between a monomer and a polymer.

39 Draw a diagram of a short section of a polypropene molecule containing three monomer units.

40 What properties of PVC ('vinyl') make it suitable for CDs and records?

41 Explain what happens when a polymer forms 'crosslinks'.

6.9 # Recycling plastics

Plastics can meet very specific technical requirements to fulfil a whole range of very different uses. They can be used for items as different as hygienic, transparent boxes to hard, rubber-like balls.

There are many benefits of plastics:
- They weigh less than other materials.
- They are extremely long lasting.
- They are resistant to chemicals, water and physical damage.
- They have excellent insulation properties.
- They are inexpensive to produce.

The problem with plastics

Most plastics will not rot away quickly like other organic matter. This is partly because there are no bacteria that naturally live on the plastics we have created. With more and more plastics being thrown away, the space required to dispose of the waste is a concern. A massive 85% of plastic waste is sent to landfill sites. Only 8% is incinerated and only 7% is recycled. Re-using plastic is even better than recycling. It reduces waste and uses less energy. Multi-trip plastic packaging has become widespread, replacing less durable and single-trip alternatives. For example, supermarkets now use returnable plastic crates for transport and display purposes. They usually last up to 20 years.

Making degradable plastics

There are now biodegradable carrier bags. These are made from plastic which rots under certain conditions or after a certain length of time. There are two types of degradable plastic:
- biodegradable plastics, which contain a small percentage of non-oil-based material;
- photodegradable plastics, which break down when exposed to sunlight.

In Sweden, McDonalds have used biodegradable cutlery since 2002. This means that all their catering waste can be turned into compost. Carriers for packs of drinks cans are also produced which photodegrade in six weeks.

There are three major concerns related to degradable plastics.
- They will only degrade if exposed to the right conditions. So, a photodegradable plastic will not degrade if it is buried in a landfill site.
- Some produce the greenhouse gas methane as they degrade.
- They may lead to an increase in plastic waste, if people believe that discarded plastics will simply disappear.

Activity – Contents of your dustbin

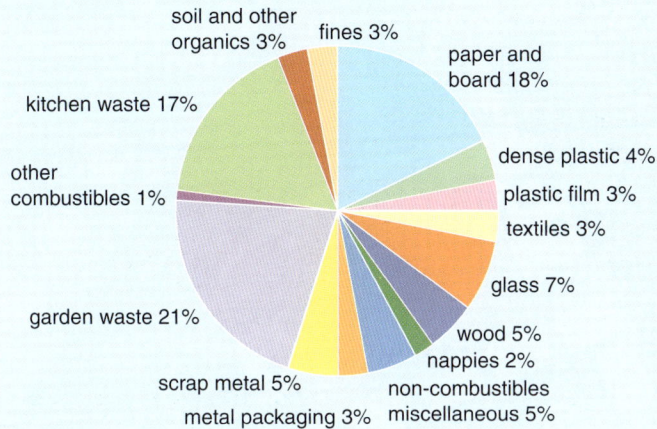

Figure 6.32 The contents of a dustbin

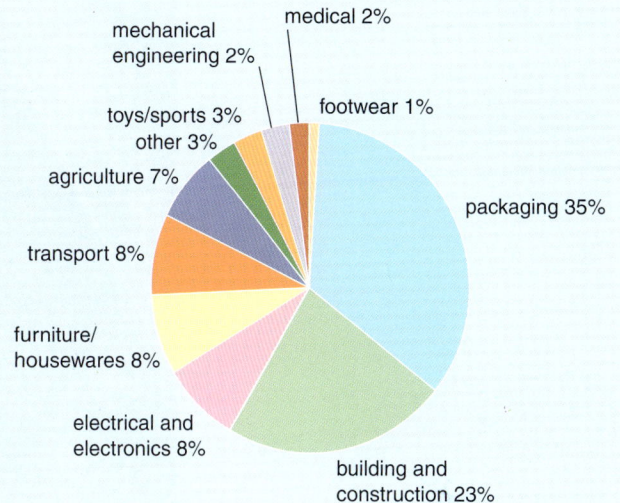

Figure 6.33 The uses of plastics in the UK

Read this statement and then answer the questions below.

'Plastic can be thought of as oil on its way to being burned. Instead of just being burnt as a fuel, the initial material from crude oil gets a life as a plastic bag, before it's burnt. When waste plastic is burnt, the energy produced can be used to generate electricity or in neighbourhood heating schemes. The large-scale burning of waste and use of the energy produced should be the priority, rather than getting every busy person to sort household waste and send it to the right place. Life in the twenty first century is just too busy and interesting for people to do that.'

1 What percentage of your rubbish is plastic?

2 a) What percentage of total plastic production goes towards making items for 'household' use?
 b) What percentage of plastic production goes towards 'job related' use?

3 Describe three ways in which recycled polythene helps the environment.

4 Are the opinions in the statement above valid? Say whether you agree or disagree with what is said and give your reasons.

Producing smart polymer materials

The rate of innovation in polymers is so fast that by the time a new advance gets into the papers, it may already be out of date. Many companies invest huge sums of money in developing new plastics.

Shape memory polymers

Doctors put a straightened plastic thread into a main artery. Once in place, the shape memory returns the plastic to a coiled spring. This widens the artery, improving the patient's blood flow.

Figure 6.34 Shape memory polymers in the casing of mobile phones are activated by heating. The phone then falls apart into its components, so that the expensive materials can be sorted and recycled. This also makes it easier to dispose of any parts containing toxic chemicals

These polymers can also be used in surgical sutures (stitches to hold flesh together). After insertion, the sutures regain their memory shape and slowly tighten, making a better repair.

Dentistry

Smart polymers can also be used in dental braces. They can be moulded into the shape of the mouth, then hardened by exposure to a special ultraviolet lamp. (The uses of smart memory metals are described in Section 5.7.)

Wound dressings

New water-based polymers can be applied to a wound. These are smart hydrogel materials. They will react to the state of the wound, sometimes hydrating it, sometimes drawing water from it. They are particularly useful for messy wounds, where there is a lot of discharge and pus. The smart hydrogel absorbs this and keeps the wound healing steadily.

Contact lenses

Ordinary contact lenses are easily scratched and can be difficult to keep clean. The wearer can also find them uncomfortable. One-day contact lenses, that you wear once and then throw away, are the answer. This is another advance brought about by silicone hydrogel polymers.

Figure 6.35 Advances in polymers have allowed us to make throw away contact lenses

Figure 6.36 Supergel for cleaning oil slicks is a hydrogel polymer. It can absorb up to half its own weight of oil. A liquid form of the gel can be sprayed on an oil slick. The gel absorbs the oil and thickens. The resulting mixture is strong and solid enough to be rolled up and lifted out of the water

Thermochromic materials

Thermochromic materials are polymers that change colour at different temperatures. They are often used on T-shirt and coffee mug designs. When they cool down, the colour changes back again.

Figure 6.37 Thermochromic materials on a coffee mug

Conducting polymers

These are really amazing materials. When a small electrical current is passed through them they contract, stretch or twist like a muscle fibre. They can be incredibly small fibres that nanorobots of the future will use as moving parts rather than electric motors.

Slime

Polyvinyl alcohol consists of long chain molecules that are free to move around in solution. When sodium tetraborate (borax) solution is added, links are made between the chains (Figure 6.38). As you add more cross-linking borax solution, the 'slime' you produce behaves more like a solid.

Figure 6.38 Polymers: two liquids make a semi-solid

Summary

✓ **Crude oil** is a mixture of hydrocarbons that can be separated into fractions of similar molecular size by fractional distillation.

✓ **Hydrocarbons** are compounds containing only carbon and hydrogen.

✓ The **alkanes** are a series of similar hydrocarbons.

✓ The heavier fractions in crude petroleum can be turned into useful materials by **catalytic cracking**.

✓ The **alkenes** are a series of hydrocarbons with one carbon–carbon double bond per molecule. They are more reactive than alkanes.

✓ **Polymers** are large molecules made by joining together smaller molecules called monomers.

✓ **Ethene** can be polymerised to form polyethene. The double bond in ethene molecules can be opened up to link the molecules together.

✓ Different polymers based on alkenes can be produced with a wide range of properties and uses.

✓ New polymers with novel uses are continually being developed.

✓ Burning fossil fuels can lead to a range of atmospheric pollution problems.

✓ Plastics create pollution problems if they are not biodegradable.

EXAM QUESTIONS

❶ a) Octane is an alkane with eight carbon atoms per molecule. What is its formula? *(1 mark)*
b) Write the formula of an alkene molecule with four carbon atoms. *(1 mark)*
c) Draw the structural formula of an alkene with four carbon atoms. *(1 mark)*

❷ How does an alkene decolorise bromine water? *(2 marks)*

❸ a) What structural feature of an alkene allows it to undergo polymerisation reactions? *(1 mark)*
b) Draw a short section of a poly(ethene) molecule showing six carbon atoms. *(1 mark)*

❹ Some fractions of crude oil were tested for viscosity (runniness) using the apparatus shown in Figure 6.39.

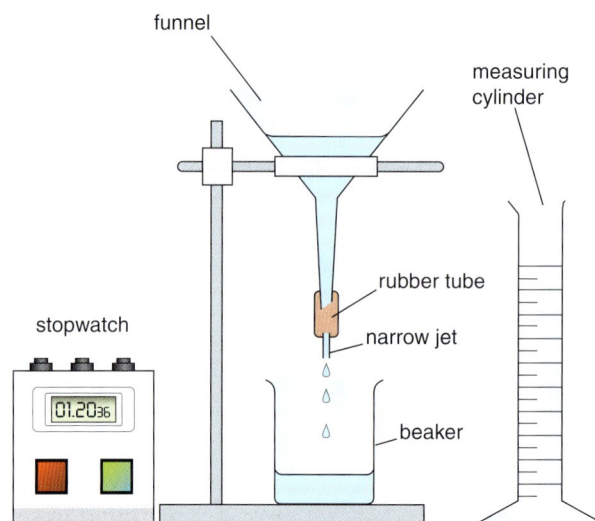

Figure 6.39

a) Explain how you would use this apparatus to compare the viscosities of fuels. *(2 marks)*
b) Name one factor that must be constant to make this a fair test. *(1 mark)*

The results of four tests are shown below.

Fraction	Time (min)
Petrol	2.0
Kerosene	2.3
Diesel	3.0
Lubricating oil	6.0

c) Copy the grid in Figure 6.40 and plot these results on to it. *(2 marks)*
d) Turn the plot of the results into a bar chart or a line graph, whichever you think is more suitable. *(1 mark)*

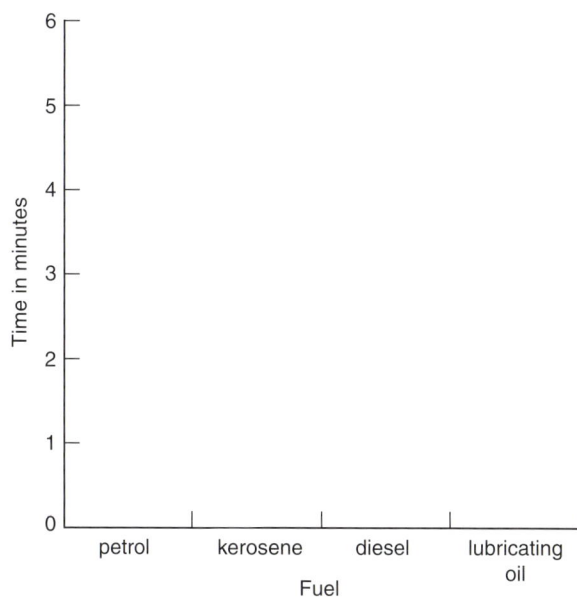

Figure 6.40

e) Explain the pattern in the results. Use what you know about the different molecular sizes in the fractions to write your answer. *(2 marks)*

❺ Latex is a naturally occurring polymer. Latex is quite soft, like a pencil eraser.

In order to manufacture the rubber for tyres, sulfur is added to latex. The sulfur makes cross-links between the polymer molecules in the latex.

a) Draw diagrams to show latex molecules before the sulfur is added and then the tyre rubber with sulfur crosslinks between the latex molecules. *(2 marks)*
b) How does sulfur change the properties of latex? *(2 marks)*

❻ Crude oil is separated by fractional distillation using a column as in Figure 6.41.
a) Explain fully how fractional distillation works. *(2 marks)*
b) Explain why naphtha burns more easily than diesel oil. *(1 mark)*
c) Naphtha contains a saturated hydrocarbon with the formula C_7H_{16}. Draw the structural formula of this compound. *(1 mark)*
d) Lubricating oil is used in a catalytic cracker. Describe what a catalytic cracker does. *(1 mark)*
e) Draw three molecules that you might obtain from cracking one molecule of $C_{15}H_{32}$. *(1 mark)*

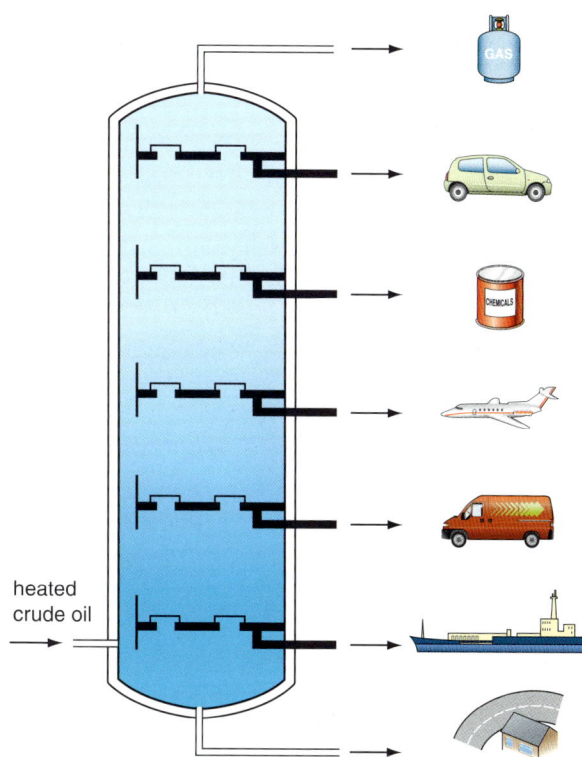

heated crude oil

Figure 6.41

Chapter 7
How can plant oils be used?

At the end of this chapter you should:

✓ know how plants provide us with plant oils;
✓ appreciate that the natural products from plants can be changed chemically to make substances that are more useful in our everyday lives;

✓ know about the extraction of plant oils to produce processed foods and biodiesel fuel;
✓ appreciate the use, benefits and drawbacks of ingredients and additives in processed foods.

Figure 7.1 Plants use raw materials in their environment. These include air, water and dissolved minerals in the ground. Animals and humans then make use of the materials that plants have made.

7.1

Plant products

Many plants contain oils. These plant oils (often called vegetable oils) are used in food and for other products, such as cosmetics and biodiesel fuel. Plant oils have been used for thousands of years for cooking and lamp oil. Years ago, the oil was extracted by pulping the plant, heating this pulp and then squeezing it in a press until the oil flowed out.

These pulping methods are still used, but more modern methods have been developed where the oil is extracted by dissolving it in a solvent such as hexane. The solvent is then distilled off to leave the oil behind.

Figure 7.2 Plants capture energy from the Sun that ends up in our food

7.2

Oil in the kitchen

Plant oil	What part of the plant does the oil come from?
Olive oil	Fruit
Rape oil	Seeds
Peanut oil	Nut (food store for seeds)
Avocado oil	Fruit
Jojoba oil	Seed
Palm oil	Fruit

Table 7.1 Different types of plant oil

Oil and water do not mix. Shake a container of oil and water for as long as you like, but as soon as you stop the oil and water will separate. The water will sink to the bottom and the oil will float to the top because it is less dense. Oil and water can be made to mix, but to do this you need an emulsifier – a substance that links together oil molecules and water molecules. Emulsifier molecules have a 'water loving' end and a 'water hating' ('oil loving') end. They bind the oil and water together forming a mixture of the two, which is called an **emulsion**.

Emulsifiers allow small droplets of one liquid to remain suspended inside the other. Egg yolk, mustard and sugars can act as emulsifiers.

❶ What raw materials do plants take in from their environment? Explain how plants obtain each of these materials.

❷ Name three products obtained from plant oil.

❸ a) What are the food groups that are essential for a healthy diet?
   b) Are there any food groups we cannot obtain from plants?

❹ Name three sources of plant oil.

❺ Old fashioned oil lamps burned plant oil. Where did the energy of the oil originally come from?

❻ Plant oils are energy stores. Explain why plant oils are mainly found in fruit (containing seeds) and in the seeds themselves.

Figure 7.3 Plant oils provide lots of energy and nutrients. They are important as foods. But beware they are very high in energy content. Eating too much plant oil-based food will make you fat.

Figure 7.4 Oil and water do not mix without an emulsifier.

Figure 7.5 Soap is a common emulsifier. When you wash your hands, the soap allows droplets of oily substances to become suspended in water.

Mayonnaise and vinaigrette

Figure 7.6 Salad dressing is made of oil and water. The thick mixture coats the salad.

To make an emulsion the oil must be broken up into tiny droplets and then prevented from coming together. Vigorous shaking or whisking breaks the oil into droplets. The emulsifier then prevents the droplets from pooling together again.

Emulsions can be temporary, like vinaigrette, or permanent, like mayonnaise. The thicker an emulsion, the less likely it is to separate, because the droplets move more slowly through the thick mixture.

Fresh milk is an emulsion of fat droplets in water. With time, the creamy fat separates from the water and floats to the top. Emulsifiers are used in many types of food including sauces, salad dressing, ice cream and cakes. In ice cream, tiny crystals of flavoured ice are trapped in droplets of cream or vegetable fat. Tiny bubbles of air are also whisked into the mixture.

Figure 7.7 Mayonnaise was invented in 1756 by the French chef of the Duc de Richelieu. After the Duc beat the British at Port Mahon, Minorca, his chef created a victory feast that included a sauce normally made of cream and eggs. As there was no cream in the kitchen, the chef used olive oil instead and a new culinary creation was born. The chef named the new sauce 'Mahonnaise' in honour of the Duc's victory.

Figure 7.8 There are many uses for emulsions, depending on their specific properties. Ice cream has tiny ice crystals suspended in oily droplets with a creamy smooth texture and delicious appearance! Gravy and salad dressings are thick so they stick to food. In fact, nearly all the sauces a chef makes are emulsions.

Frying in oil

Frying is a fast cooking method. Plant oils can be heated up to 230 °C before they start smoking, but water can only be heated to 100 °C. Direct contact between the food and the oil transfers energy rapidly to the food, causing chemical changes. You should always fry food with the oil above 100 °C. This seals in the flavour and also prevents oil soaking into the food as steam escapes from the food to stop this happening. Frying adds flavour. The high temperature changes proteins and sugars on the surface so they turn brown and taste good. You should always take care when helping to prepare fried food. In the UK alone there are 3000 injuries from chip pan fires every year!

Figure 7.9 Frying adds to the flavour of food

7 Draw step-by-step diagrams to show how olive oil and lemon juice (a watery solution) can be used to make mayonnaise. Egg white is used as the emulsifier. The oil must be added a little at a time.

8 Why are emulsions used as sauces and salad dressings?

9 Why do chips cook faster than boiling potatoes?

10 Why do chips taste different from boiled potatoes?

7.3 # Food additives

Adding chemicals to food isn't a recent invention. Saltpetre was used in the Middle Ages to preserve meat. Saltpetre is the common name for potassium nitrate. Nowadays, nitrate, the active ingredient in saltpetre, is used. It prevents meat from becoming contaminated with the bacteria that cause food poisoning.

Additives are used to make food look and taste more attractive, and to prevent it from turning mouldy or stale. But, there is a concern that food additives make unhealthy, processed foods cheaper and more attractive than healthy, fresh foods.

Why are additives given E-numbers?

The European Union (EU) requires food additives to be labelled clearly in the list of ingredients, either by name or by E-numbers.

Figure 7.10 Lots of foods contain additives

If an additive has an E-number, it means that it has passed safety tests.

What types of additives are there?

There are many different types of food additives which are grouped together by what they do to food. Additives include:

- antioxidants, which keep food fresh;
- colourants;
- emulsifiers, stabilisers, and thickeners;
- flavourings;
- preservatives;
- sweeteners.

The use of additives can be beneficial.

- Antioxidants stop butter being oxidised and going rancid if it is left out of the fridge for too long.
- Some additives prevent bread going mouldy.
- Preservatives kill the bacteria that produce deadly poisons in meat.
- A teaspoonful of high-intensity sweetener has the same sweetening effect as a kilogram of sugar and yet it is calorie-free and virtually harmless to teeth.

Substances present in fresh tomatoes

Flavouring:
- E621 monosodium glutamate

Colourants:
- E160a carotene
- E160d lycopene
- E101 riboflavin (vitamin B6)

Antioxidant:
- E300 ascorbic acid (vitamin C)

Acids:
- E330 citric acid
- E296 malic acid
- Oxalic acid

Figure 7.11 Most additives occur naturally in food, they are not manufactured. All the substances listed on the left with E-numbers are present in a fresh home-grown tomato. They have not been added, they are in the tomato naturally. Sometimes these substances are added to other foods to improve them.

Additives can also have drawbacks.

- Aspartame, a sweetener, is widely used in soft drinks. When aspartame was fed to rats, it resulted in low levels of tryptophan in the brain. Low tryptophan levels are linked with aggressive and violent behaviour.
- Tartrazine (E102), the orange colour in some soft drinks, has been linked to asthma, rhinitis and hyperactivity.
- Sunset Yellow (E110), used in biscuits, has been found to damage the kidneys.

Analysing food additives

Chemists use liquid chromatograms to test foods for additives. Their machines use the same chromatography principles as separating ink colours with water on a filter paper chromatogram. In the machine, the paper is replaced by a column of granules and, instead of water, the liquid is a mixture of solvents. After passing through the chromatography machine, the substances are analysed by a mass spectrometer. This machine helps to identify the separated substances.

⑪ Explain simply how food can be tested for artificial colours.

⑫ a) What are E-numbers?
   b) Why are they used?

⑬ Give two benefits of using food additives.

⑭ Give two drawbacks of using food additives.

⑮ Why are food additives blamed for making unhealthy food more attractive?

Figure 7.12 You've seen simple colour chromatography – analytical chromatography follows the same principles

7.4

The marge story

Have you noticed the pictures of sunflowers on margarine tubs? But, what's the connection between sunflowers and margarine?

Sunflower oil is extracted from sunflower seeds. The seeds are crushed between heavy rollers to squeeze out the oil and this is used to make margarine. Sometimes the seeds are heated to obtain a higher yield of the oil, but this can lead to lower quality, more acidic oils.

Olive oil is also used to make 'marge'. There are several grades of olive oil. Extra virgin olive oil is produced by pressing the flesh of ripe olives at room temperature. This gives the highest possible quality of olive oil. Ripe olives yield about $200\,cm^3$ of oil per kg of fruit. Commercial growers can get more oil out of the olives, but this has a higher acidity and is used for making soap.

Butter is made from the cream in milk. This is an animal fat. Unfortunately, animal fats contain cholesterol which causes heart disease. Margarine is a butter substitute. You can remove a lot of the cholesterol from your diet by eating margarine instead of butter.

Liquid plant oils, like sunflower oil and olive oil, can be changed into solid margarine by hydrogenation. This means that hydrogen is added to the oil molecules.

Figure 7.13 Sunflowers are a common crop in Southern Europe

Plant oils contain **unsaturated** molecules with carbon–carbon double bonds. Molecules with these double bonds tend to be liquids (oils) at room temperature. If you add hydrogen to the double bond, the liquid oil turns into a much more solid fatty material which is easy to spread on bread.

The liquid oil is heated to 60 °C with a powdered nickel catalyst and hydrogen gas is bubbled through it. When the oil is filtered and cooled it turns into a solid.

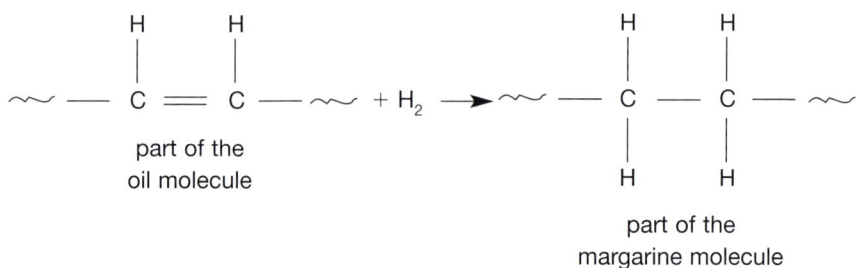

Plant oils belong to a family of chemicals called triglycerides (Figure 7.14). They are not the same as crude oil, which contains hydrocarbons. The shape of a plant oil molecule indicates how useful it is.

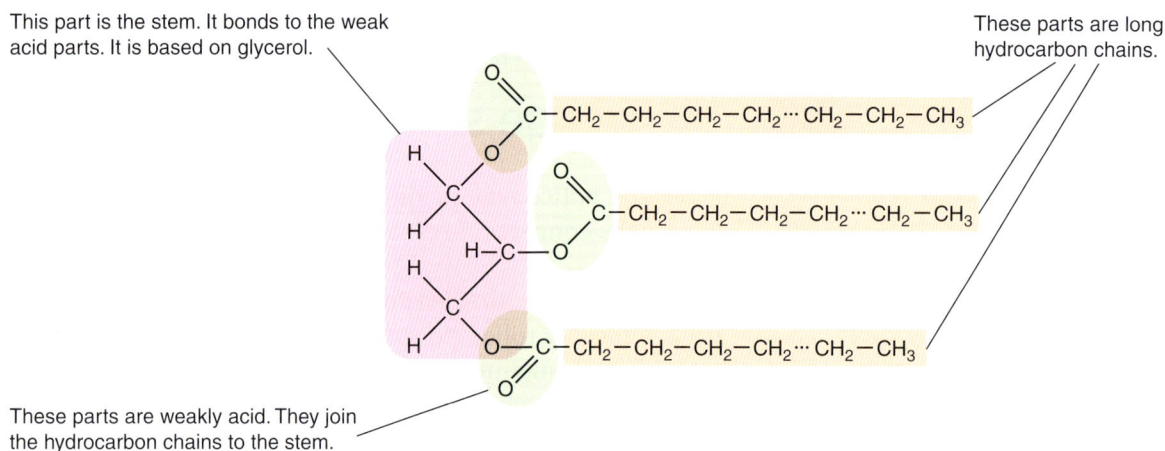

Figure 7.14 The structure of a typical plant oil molecule

⑯ One kilogram of olives produces 200 cm^3 of oil. Olive oil has a density of 0.9 g per cm^3. What percentage of the olives is oil?

⑰ Why could eating a lot of olives make you fat?

⑱ Why is extra virgin olive oil the best olive oil for cooking?

⑲ Draw a flow chart to summarise the process of changing sunflower oil into margarine.

⑳ What is a triglyceride?

7.5 Polyunsaturated fat

An **unsaturated** fat is a fat which contains one or more carbon–carbon double bonds in its molecules. Such fat molecules are **mono-unsaturated** if they contain one double bond and **polyunsaturated** if they contain more than one double bond. See also page 108.

An unsaturated fat has chains like this:

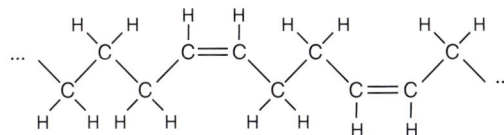

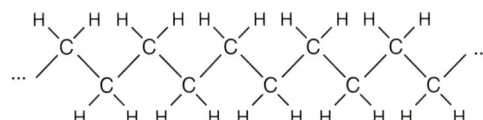

A saturated fat has chains like this:

Figure 7.15 The structure of molecules of saturated and unsaturated fats

Saturated fats are present in meat and dairy products such as milk, cheese and cream. When margarine or cooking fat is made from corn oil, soyabean oil or other plant oils, hydrogen atoms are added producing **saturated** fat molecules and consequently the oil becomes more solid.

When we reduce the amount of saturated fat in our diet, it reduces both our blood cholesterol level and our chances of developing heart disease. It is therefore important to read food labels and check how much saturated fat is in our food (see also page 21).

Unsaturated fats are usually liquid at room temperature. They are found in most plant oils. Choosing foods containing polyunsaturated fats can reduce your risk of heart disease. But, fat is fattening, and unsaturated fats pile on the kilojoules (calories) of energy, just like saturated fats. So, eating too much unsaturated fat will increase your weight just like saturated fat.

Testing an oil or fat to see if it is unsaturated

1 Place a few drops of oil or melted fat in a test tube.
2 Put the test tube in a 250 cm³ beaker of hot water (Figure 7.16).
3 Add three drops of bromine water, or 2% iodine solution.
4 Stir briefly.
5 Time how long it takes for the red or orange colour to fade away.

Bromine and iodine react with any unsaturated carbon–carbon double bonds in the oil or fat molecule. This reaction forms a colourless compound. If the colour fades away in a few seconds the fat or oil is high in unsaturated fat.

Bromine water is an irritant and eye protection should be worn during this test.

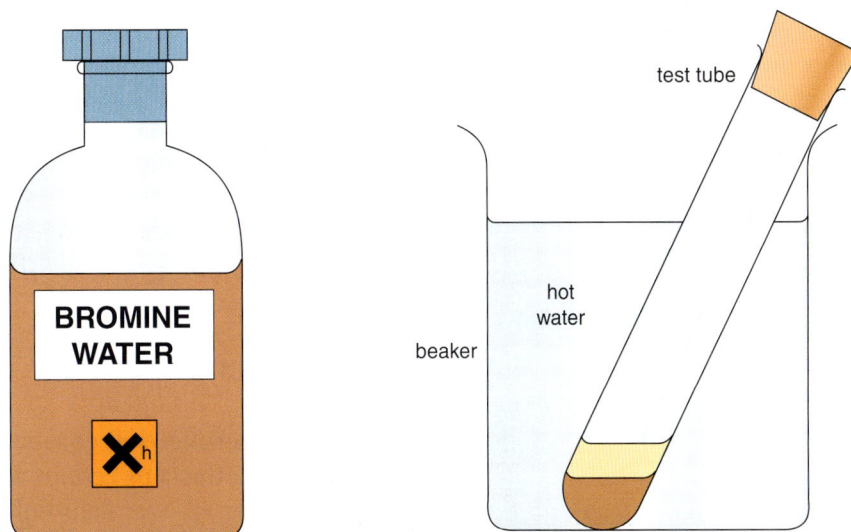

Figure 7.16 Test for unsaturated fats

Here are some tips to help you reduce fat in your diet:
- eat less fatty foods;
- eat more low-fat foods such as fruit, vegetables, bread, rice, pasta and cereals;
- use less fat, less oil, less butter and margarine in cooking;
- use skimmed milk and low-fat cheeses instead of whole milk and cheese;
- choose margarine instead of butter;
- trim the fat off meat and remove the skin from poultry;
- read and compare food labels to find foods that have less total fat.

21 Why is saturated fat bad for you?

22 Why is eating too much unsaturated fat also bad for you?

23 Draw an unsaturated fat molecule and write a caption to explain why it is unsaturated.

24 How would you test olive oil to see if it was an unsaturated fat?

7.6 Do we need so much oilseed rape?

Figure 7.17 Yellow flowers of oilseed rape. The name 'rape' is derived from the Old English word for turnip, *rapum*

25 Describe an oilseed rape plant. What would you see and smell?

26 Name three sources of plant oil.

27 How is oil extracted from rape seeds?

28 a) What use is made of the material left after the oil is extracted?
   b) Why is this material so useful?

Oilseed rape is a very useful crop. Its seeds are 42% oil and the material (meal) left after removing the oil is 42% protein and very good as animal food.

Rape seeds are tiny round black seeds which germinate rapidly. After a period of growth, the rape plant produces bright yellow flowers, smelling faintly of honey.

Oilseed rape is widely cultivated throughout the world for the production of animal feed, plant oil and biodiesel. Forty million tonnes of seeds are produced worldwide every year.

Oilseed rape is the third leading source of plant oil in the world after soyabean oil and palm oil. It is also the world's second leading source of protein meal for animals.

After storage, the seeds are crushed between rollers, to extract the oil. There is so much oil in the seeds that other extraction methods are unnecessary.

GM controversy

Some seed companies sell genetically modified (GM) rape seeds that are resistant to the effects of certain weedkillers. This allows farmers to spray weedkiller all over their fields without affecting the growth of their GM oilseed rape crop.

29 Name a non-food use of rapeseed oil.

30 What are the drawbacks of cultivating rape seed?

31 On balance, do you think the cultivation of oilseed rape should be increased or decreased? Explain your decision.

In recent years, these companies have prosecuted farmers found to be growing their seeds without paying for them. The farmers say that GM pollen was carried to their fields and combined with the non-GM rapeseed which they had planted.

Other farmers find unwanted GM plants in their fields that are not killed by weedkiller.

The extensive use of weedkillers when farming these GM crops leads to a significant loss of biodiversity, as wildflowers ('weeds') are killed, leaving wildlife, such as bees and butterflies which are dependent on the wildflowers, unable to survive.

Biodiesel

Rapeseed oil can be turned into diesel fuel by chemically pulling the plant oil molecules apart. This fuel is 'carbon neutral' because the carbon dioxide from the burning fuel was removed from the air by the rapeseed plants when they were growing.

Petroleum-based fuel is running out and becoming very expensive. This has led to the suggestion that locally produced motor fuel could become one of the fastest-growing industries in the future. Small-scale factories could spring up in every town and city to produce biodiesel. You could become a multi-millionaire working in this industry!

7.7 Alternatives to petrol and diesel

Figure 7.18 H₂ Eyes Cool is hi-tech. She sells photovoltaic cells that turn water into hydrogen fuel and converts your car to run on it. Sunshine, rainwater and some technology are all you need. She's light and bright

Activity – What happens when petrol runs out?

In a few years petrol will be running out.

Locally produced motor fuel has become the fastest growing industry and small-scale factories are springing up in every town and city producing alternatives to petrol and diesel. You have to decide which fuel you will support.

Work in groups for this activity.

H₂ Eyes Cool is selling hydrogen and fuel cells

- The raw materials for this form of energy are sunlight and moisture in the air.
- Photovoltaic (solar) cells are becoming common and more efficient. They split up water forming hydrogen for the fuel cells and oxygen, which is vented to the air.
- There will never be a shortage of fuel, as long as the Sun shines.
- Photovoltaic cells are expensive.

- The cells will only make hydrogen fuel in daylight hours.
- Hydrogen fuel is clean – the waste materials from the process are oxygen and water.
- Storing hydrogen is difficult, but the problem has been solved using carbon nanotubes. They can store 65% of their mass as hydrogen.
- Hydrogen cars don't have an engine that burns fuel, so noisy exhausts will be a thing of the past.
- Hydrogen cars have an electric motor. The hydrogen powers a fuel cell to make electricity. The oxygen you need comes from the air.
- Hydrogen car technology is expensive.

Betty Biodiesel
- You don't have to modify your ordinary diesel engine to burn biodiesel.
- Biodiesel can be mixed with ordinary diesel to make it last longer.
- Biodiesel contains 80% less CO_2 and does not produce SO_2, so acid rain will be reduced.
- Biodiesel leaves 90% less unburnt hydrocarbons to make ozone and smog.
- Biodiesel produces lubrication, so engines last longer.
- Biodiesel has been used successfully in Europe for 20 years.
- Biodiesel is safe to handle and transport. It does not burst into flames.
- Biodiesel is made from crops such as soy, sunflowers and rape seed.
- Biodiesel is as biodegradable as sugar and ten times less toxic than salt.
- Biodiesel exhaust fumes smell of popcorn.

Sugar Dude – Ethanol fuel
- Ethanol is made by the natural process of fruit rotting.
- Crops such as maize, barley and potatoes can be used instead of fruit.
- The fuel is 'carbon neutral' because the original crops took in carbon dioxide.
- Ethanol burns very cleanly producing no polluting gases such as sulfur dioxide.
- Ethanol is not a hydrocarbon, so there can be no unburnt hydrocarbons to cause smog.
- Ethanol mixes with water and does not pollute it.
- Making ethanol fuel is a two stage process – fermentation and distillation.
- Ethanol is very flammable and bursts into flames easily.
- Pure carbon dioxide is produced by fermentation. This can be collected and used to make fizzy drinks.
- After fermentation the 'mash' can be made into animal feed.
- Ethanol keeps your engine clear and clean.

Figure 7.19 Betty Biodiesel makes diesel motor fuel from recycled cooking oil and plant oil from local farmers. She's keen and green

Figure 7.20 Sugar Dude runs a fermenting plant. She makes ethanol fuel for motor vehicles. She's cooking up a clean future

❶ Compare each of the alternative fuels with petrol and diesel. Focus on the disadvantages of the 'new fuel'.

❷ Make a table to compare the benefits and drawbacks of each fuel.

❸ Appoint one person to make a short (3-minute) presentation for *each* fuel.

❹ Decide as a group which fuel is best. You will need to decide on criteria such as cost, cleanliness, availability and supply.

❺ Devise a plan for setting up a small factory to produce your chosen fuel for the petrol stations in a small town.

Summary

✓ Plants can provide oil from seeds and fruits.

✓ Mixtures of oil and water are called **emulsions** and are widely used in cooking.

✓ **Frying** in oil cooks food faster than boiling and changes the flavour.

✓ **Additives** are chemicals used to improve food, but they can have drawbacks.

✓ Plant oils can be made into margarine, an alternative to butter.

✓ Unsaturated fats from plant oils are more healthy than saturated animal fats. The most healthy diet is low in all fats.

✓ Plant oil crops can be made into a motor fuel called biodiesel.

EXAM QUESTIONS

❶ a) Where do plants get their energy from? *(1 mark)*

b) Copy and complete these sentences.
A plant collects energy using _____ .
This makes _____ in the leaves of the plant. Plants also store energy as _____ in their seeds. This energy store is to help the seed _____ . *(4 marks)*

❷ Food with additives looks better, tastes stronger and lasts longer on the shelf. Who benefits most from this – the manufacturer, the retailer or the customer who eats the food? Explain your answer. *(3 marks)*

❸ Write a 50-word report to be read on a news programme, warning people about saturated fats.

The message should be amusing to catch the attention of the audience. *(4 marks)*

❹ This question is about homemade mayonnaise.

Homemade mayonnaise
Ingredients
1 egg yolk
150 cm^3 olive oil
10 cm^3 grainy mustard
50 cm^3 fresh lemon juice
salt and freshly ground black pepper to taste

Method
1 Place the egg yolk in a large bowl.
2 Gradually whisk the oil into the egg yolk.
3 When the egg and oil are well combined, whisk in the mustard and lemon juice.
4 Season the mayonnaise.

a) What plant oil is used to make the mayonnaise above? *(1 mark)*
b) What is the watery solution in the mayonnaise? *(1 mark)*
c) Which ingredients act as emulsifiers? *(1 mark)*
d) How do the emulsifiers prevent the oil and water from separating out? *(1 mark)*

Chapter 8
What changes have occurred in the Earth and its atmosphere?

At the end of this chapter you should:

✓ know that the Earth has a layered structure comprising a crust, a mantle and a core;

✓ appreciate that the Earth's crust and the upper part of the mantle are cracked into a number of very large tectonic plates;

✓ understand how the slow movement of these tectonic plates can cause earthquakes, volcanic eruptions and changes in the landscape;

✓ understand why scientists cannot predict accurately when earthquakes and volcanic eruptions will occur;

✓ know about the composition of the Earth's atmosphere;

✓ be able to explain the changes that have occurred and are continuing to occur in the Earth's atmosphere.

Figure 8.1 The Earth from space – a blue and white planet. Can you see continents, oceans and swirls of white clouds which form part of the atmosphere? In this chapter we shall be studying the Earth and its atmosphere.

8.1

The Earth and its atmosphere

In the last three chapters, we have seen that the Earth and its atmosphere provide the raw materials for everything we need:

- the Earth itself provides crude oil, rocks for building materials and metal ores;
- the seas provide salt (sodium chloride) from which we obtain sodium, chlorine and sodium hydroxide;
- the atmosphere provides important gases including oxygen and nitrogen;
- living things, animals and particularly plants provide foods and raw materials for clothing and medicines.

1 Write the following steps in the order they occur when oxygen is separated from argon and nitrogen in the atmosphere:
A allow liquid air to warm up
B compress and cool air
C nitrogen boils off first, then argon, then oxygen
D oxygen liquefies first, then argon, then nitrogen.

Figure 8.2 A variety of people require extra supplies of oxygen to survive. These include hospital patients, victims of accidents, divers and mountaineers. Pure oxygen can be separated from other gases in the atmosphere by fractional distillation of liquid air

8.2

Layers of the Earth

The Earth is a huge ball of rock, iron and nickel. It has a radius of about 6400 km. When the Earth was first formed, about 4500 million years ago, it was a mass of molten rock and metals. Over millions of years, the Earth cooled down. At the same time, heavier materials sank towards the centre of the Earth, less dense materials rose to the surface and this resulted in a layered structure (Figure 8.3).

There are three distinct layers in the Earth:
- a thin outer **crust** of less dense material on the surface of the Earth about 50 km thick. In places where the crust is thicker, its surface is above sea level.
- a **mantle** of moderately dense rock about 3000 km thick that extends almost halfway to the centre of the Earth. Temperatures in the mantle range from 1500 to 4000 °C. The hottest rocks in the mantle form a very viscous liquid called **magma**, which changes shape and flows very slowly.
- a **core** of very dense iron and nickel with a radius of 3500 km (about half the Earth's radius).

Most of the evidence for the Earth's layered structure comes from the study of shock waves from earthquakes. Some of the waves are reflected by layers of the Earth and this reflection shows up the crust, the mantle and the core very clearly.

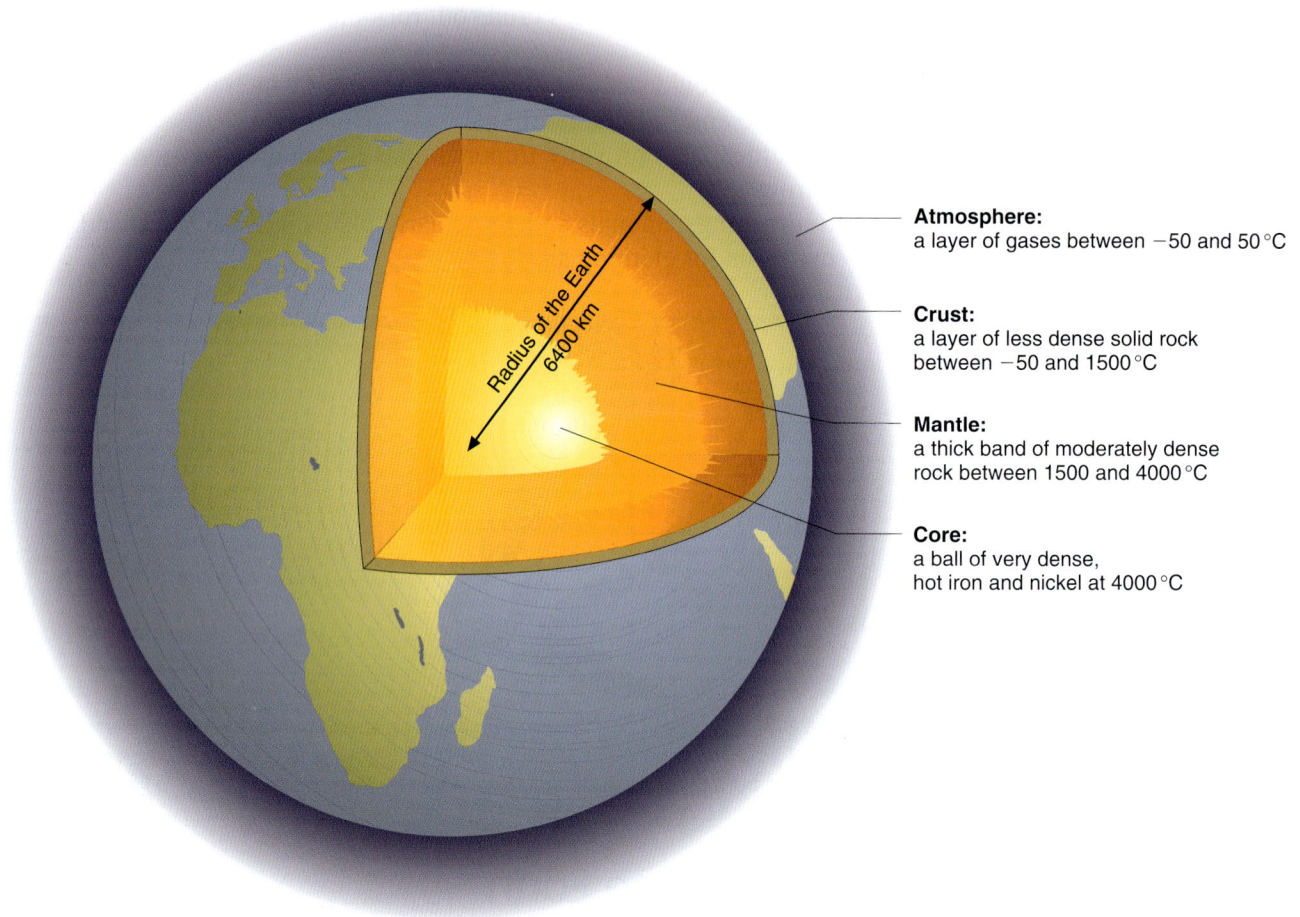

Atmosphere:
a layer of gases between −50 and 50 °C

Crust:
a layer of less dense solid rock between −50 and 1500 °C

Mantle:
a thick band of moderately dense rock between 1500 and 4000 °C

Core:
a ball of very dense, hot iron and nickel at 4000 °C

Radius of the Earth 6400 km

Figure 8.3 The layers of the Earth

❷ a) How do the temperatures of the different layers in the Earth change towards the centre?

b) How do the densities of the layers in the Earth change towards the centre?

More detailed studies have suggested that the core can be split into two – an **outer core**, which is liquid, and an **inner core**, which is solid.

Outside and above the Earth there is a fourth layer – the **atmosphere**, which is a layer of different gases about 100 km thick.

Why is the Earth's core so hot?

One theory about the origin of the Earth is shown in Figure 8.4.

Although the crust and mantle insulate the Earth's core, there is a second, more important, effect that helps to maintain high temperatures inside the Earth. Some rocks in the Earth, particularly granite, contain radioactive atoms of elements such as uranium and potassium (see Section 11.8). As the nuclei of these elements break up (decay), energy is released as heat and electromagnetic radiation. This energy helps to maintain temperatures inside the Earth.

❸ The theory summarised in Figure 8.4 suggests that the Earth began as a giant molten ball. As this cooled down, the crust and other layers formed.

a) Why did iron and nickel move into the core when layers formed in the Earth?

b) Volcanoes are responsible for some features on the surface of the Earth. Suggest two ways in which other features, such as valleys and mountains, formed on the Earth's surface.

c) In the last few years, average temperatures on the surface of the Earth have risen slightly. Why do you think this has happened?

d) Do you think that temperatures on the Earth will continue to rise or will they eventually fall? Explain your answer.

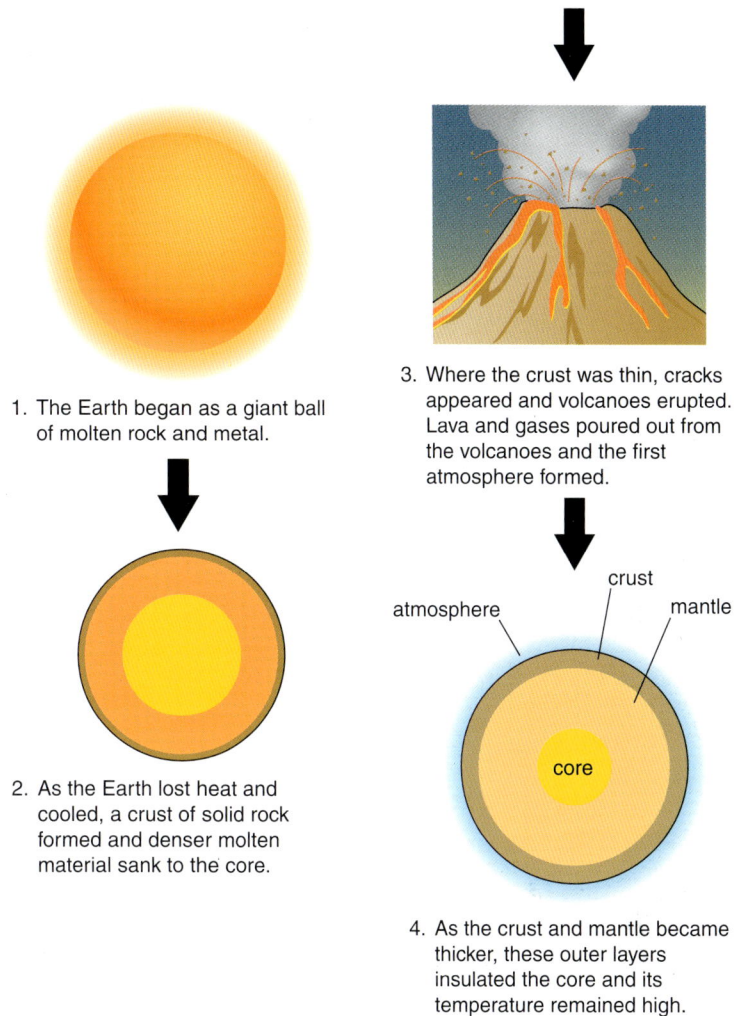

1. The Earth began as a giant ball of molten rock and metal.

2. As the Earth lost heat and cooled, a crust of solid rock formed and denser molten material sank to the core.

3. Where the crust was thin, cracks appeared and volcanoes erupted. Lava and gases poured out from the volcanoes and the first atmosphere formed.

4. As the crust and mantle became thicker, these outer layers insulated the core and its temperature remained high.

Figure 8.4 A theory to explain the high temperatures inside the Earth

8.3 How is the Earth changing?

Before 1915, most scientists thought that:
- the Earth contracted and shrunk as it cooled down;
- the shrinking caused the Earth's crust to wrinkle (Figure 8.5);
- the shrinking and wrinkling of the Earth's crust created mountains, valleys and other features on the Earth's surface;
- the hard, dense solid rocks in the Earth's crust prevented any movement of the continents.

In 1915, the German meteorologist (weatherman) Alfred Wegener suggested that the continents were in fact moving. At first, people laughed at his ideas because they seemed impossible. But, as people listened to Wegener's evidence, his ideas didn't seem quite so impossible after all.

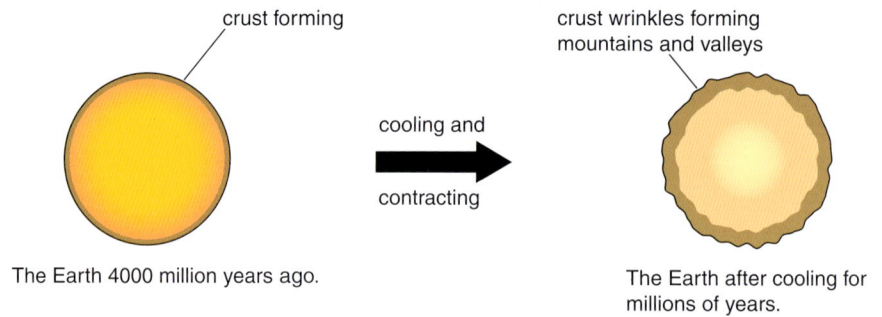

crust forming

crust wrinkles forming
mountains and valleys

cooling and

contracting

The Earth 4000 million years ago.

The Earth after cooling for
millions of years.

Figure 8.5 Early ideas about the formation of mountains on the Earth's surface

Activity – Wegener's evidence for moving continents

Continental fit

Wegener noticed that the continents looked like pieces of a jigsaw puzzle that could fit together. For example, it was easy to see that South America might fit into the coastline of West Africa (Figure 8.6).

❶ Use tracing paper to trace the outline of South America in Figure 8.6. Carefully, cut out your outline. How well does it fit into the coastline of West Africa?

❷ Most geographers and geologists thought that this was just a coincidence. How on Earth could continents move sideways by thousands of kilometres! What do you think?

Mountain chains

Wegener also noticed that mountains with similar rock types and structures occurred in different continents, but these mountain chains came close

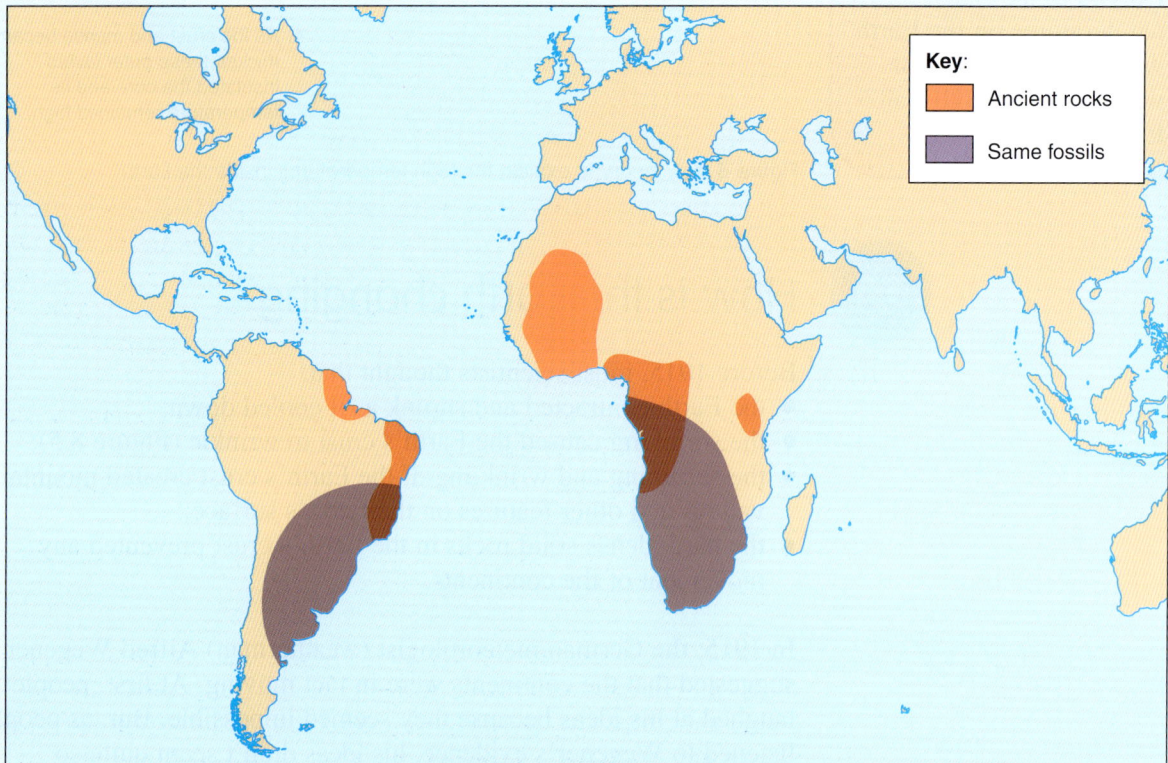

Key:
Ancient rocks
Same fossils

Figure 8.6 Wegener's evidence for the movement of South America and Africa

together when the continents formed one giant jigsaw (Figure 8.6).

❸ Shade the areas with ancient rocks on your outline of South America. Fit your outline as snugly as possible into the west coast of Africa. How neatly do the areas with ancient rocks in South America and West Africa align?

❹ Experts told Wegener that the alignment of these ancient rocks was not exact. Is that right?

Fossil records

Animals and plants can only survive in areas where the climate and habitat suits them. Yet Wegener noticed that fossils of the same animals and plants had been discovered on different continents, sometimes with different climates and habitats. Just like his observation of mountain chains, these areas with the same fossils were close together when his 'giant jigsaw' was completed.

❺ Shade the area with the same fossils on your outline of South America. Now fit your outline again into the west coast of Africa. How well do the areas with the same fossils in South America and West Africa align?

❻ Those who disagreed with Wegener's ideas suggested there was probably once a land bridge connecting South America and West Africa. Is this a possibility?

From observations similar to those you have just made, Wegener suggested that:

- about 200 million years ago, all the continents were joined together as one supercontinent which he called **Pangaea** from the Greek words meaning 'all lands';
- then Pangaea broke up and over millions of years the continents slowly moved apart. Wegener called this movement **continental drift**.

❼ How do you think Wegener was able to date the break up of Pangaea to 200 million years ago?

❽ Why do you think that Wegener's theory of crustal movement (continental drift) was not accepted for many years?

8.4 Why should the continents move apart?

In spite of Wegener's evidence, his ideas were not accepted. This was mainly because he could not explain how or why the continents should move apart. But, just before Wegener died in 1930, it was suggested that the continents could be moved by **convection currents** in the mantle. More definite evidence for these convection currents and for continental drift wasn't found until the 1950s. During the 1950s and 1960s, the floor of the Atlantic Ocean was surveyed and studied in detail. This led to some startling discoveries.

- Running down the centre of the Atlantic Ocean there is an undersea mountain ridge with volcanoes (Figure 8.7).
- On either side of this volcanic ridge there is a similar pattern of humps and hollows.
- The rock at the top of the volcanic ridge is almost new and the further rocks are from the ridge, the older they get.

All this evidence suggested that lava was spewing out from volcanoes onto both sides of the mid-Atlantic Ridge as the result of convection currents in the mantle. It looks as though the bottom of the Atlantic Ocean has been made on a stop/start volcanic production line over millions of years.

The final proof of Wegener's ideas about moving continents came in 1963.

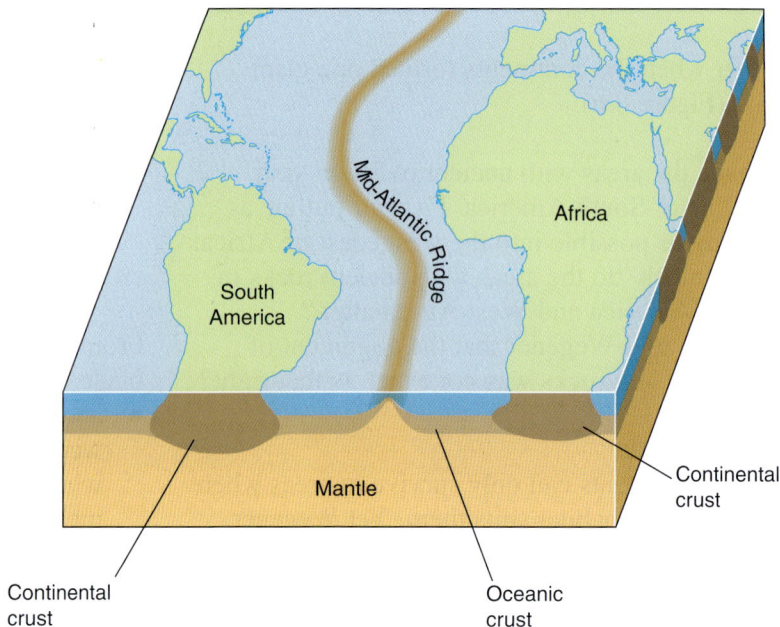

Figure 8.7 The mid-Atlantic Ridge

During the 1950s, geologists began to study the magnetism in rocks. As molten rocks solidify, any bits of iron in them become magnetised in line with the Earth's magnetic field. To their surprise, the geologists found that every half-million years or so the Earth's magnetic field flips – the North Pole becomes a South Pole and vice versa.

Then, in 1963, magnetic surveys along the mid-Atlantic Ridge showed that the magnetism of rocks on one side of the ridge was an exact reflection of the other side (Figure 8.8). Rocks now hundreds of kilometres apart must have been formed at the same time, as magma spewed out of the volcano. These rocks were pushed further and further apart over millions, if not billions, of years as new rock formed. And as these rocks moved apart, the continents of South America and Africa moved with them.

Wegener's ideas about moving continents resulted in the study of **plate tectonics**.

8.5 How are the continents moving?

The Earth's *shape* is like an orange – spherical but slightly flattened at the poles. Its *structure* is like a badly cracked egg. The 'cracked' shell is like the Earth's thin crust, the 'egg white' is the mantle and the 'yolk' is the core.

The crust and the solid upper parts of the mantle are sometimes called the Earth's lithosphere. The lithosphere is not one continuous solid shell. It is cracked and broken into a number of massive pieces called **tectonic plates**.

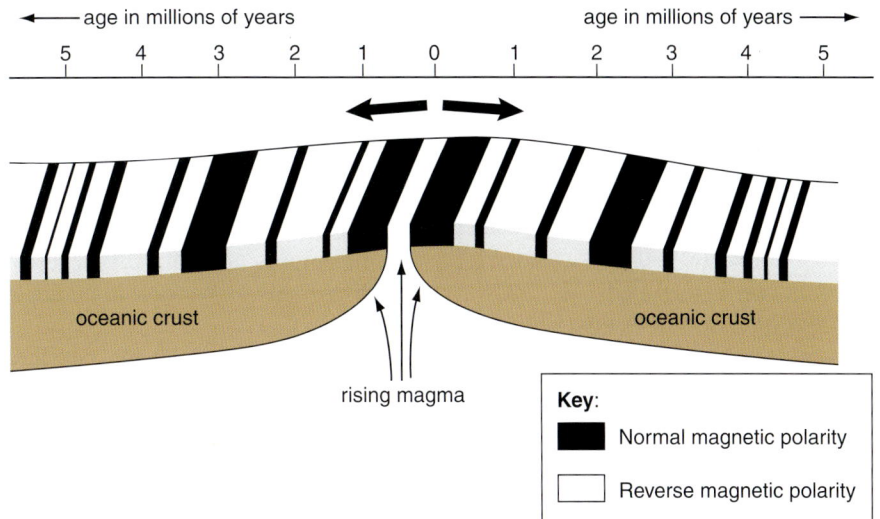

Figure 8.8 *Magnetic patterns in the rocks on both sides of the mid-Atlantic Ridge*

The Earth's core is incredibly hot, 4000 °C and higher, due to the heat released by natural radioactive processes. This causes slow convection currents in the liquid mantle and slow movements of a few centimetres per year of the tectonic plates. Figure 8.9 shows a map of the world with the main tectonic plates and their direction of movement.

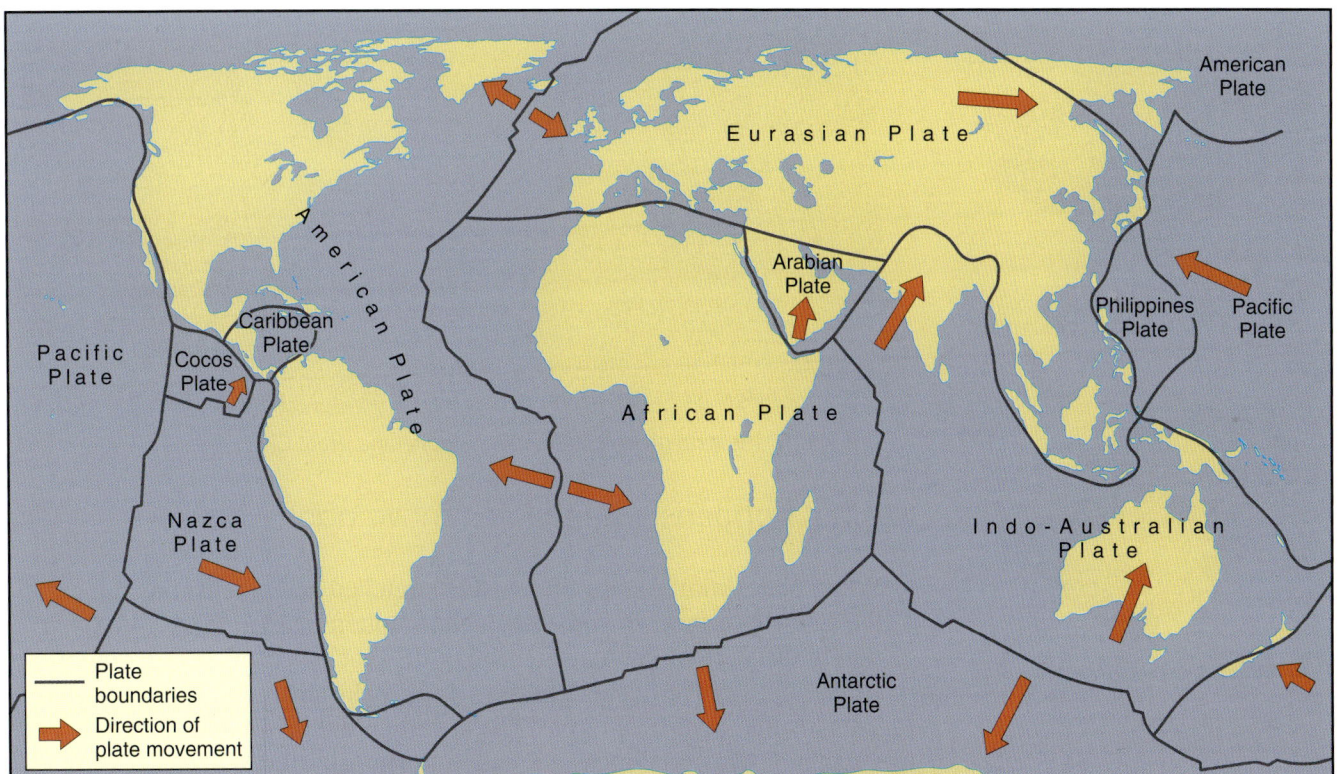

Figure 8.9 A map of the world showing the main tectonic plates and their directions of movement

Plate tectonics explain many of the features on the Earth's surface including volcanoes, earthquakes and even the formation of mountain ranges.

To appreciate the importance of plate tectonics, let's look at what happens when plates slide past each other, when they move apart and when they move towards each other. Although movements of the tectonic plates are normally very slow, they can sometimes be sudden and disastrous.

What happens when plates slide past each other?

When two plates slide past each other, stresses and strains build up in the Earth's crust. Massive forces are involved, due to convection currents in the mantle pushing tectonic plates one way or another. The forces are so great that they can cause the plates to bend. In some cases, the stresses and strains build up and are then suddenly released. The Earth moves, the ground shakes violently in an **earthquake** and breaks appear in the ground (Figure 8.10).

These breaks in the ground when plates slide past each other horizontally are called **tear faults** (Figure 8.10b)). The San Andreas fault in California and the Great Glen fault in Scotland are examples of tear faults. In fact, the map of Scotland would look very different if the Great Glen tear fault did not exist (Figure 8.11).

Measurements from space satellites have shown that the American Plate is moving away from the Eurasian and African Plates at the speed of 5 cm per year. Over millions of years, movements such as this can cause massive changes on the Earth.

❹ How many *centimetres* will the American plate have moved away from the Eurasian and African Plates in one million years?

❺ How many *kilometres* will it have moved away in one million years?

❻ At this rate of separation, how much wider will the Atlantic Ocean be in one million years?

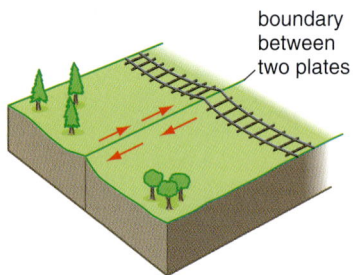

a) Plates in the Earth's crust are bent as they slide past each other.

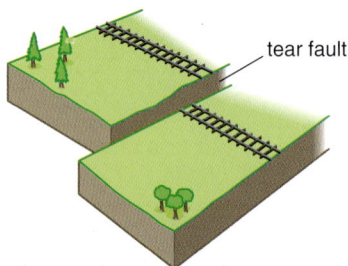

b) Stresses in the bent plates are suddenly released, the earth moves and the ground shakes violently in an earthquake. Breaks appear in the ground and a tear fault has formed.

Figure 8.10 How an earthquake occurs

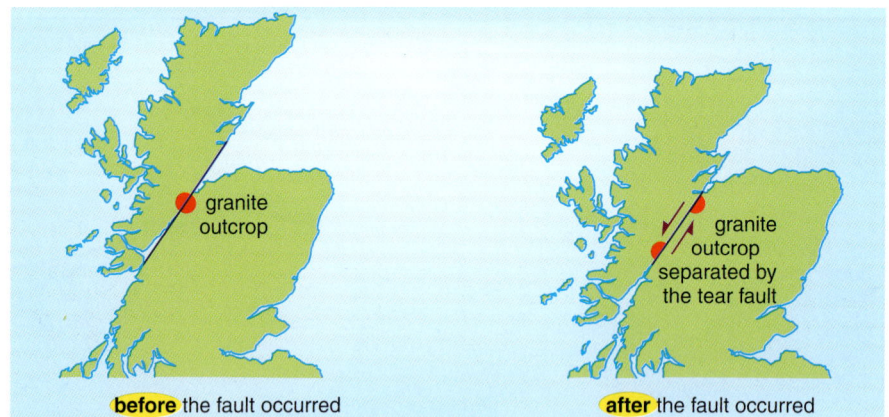

Figure 8.11 The map of Scotland before and after the Great Glen Fault

What happens when plates move apart?

When plates move apart, the crust is stretched and cracks may appear in the Earth's surface. In some cases, hot molten magma rises from deep within the Earth and escapes through the cracks, erupting as a **volcano**. If the plates move further apart, surface rocks sink and may get buried. This results in vertical faults. These vertical faults produced by stretching (tension) forces are called **normal faults**. When two vertical faults occur alongside each other, a rift valley is formed (Figure 8.12).

Rift
valley

land
slips
down

Normal fault

Normal fault

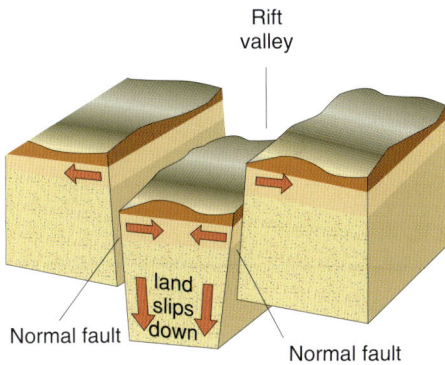

Figure 8.12 As plates move apart, the land may sink into the crack. If there are two normal (vertical) faults near each other, a rift valley may form.

❼ In California, most of the orange groves have trees growing in straight lines. In some groves, the lines of trees are kinked, but they were not planted this way. Why are the lines of trees now kinked?

In many cases, volcanoes erupt without causing too much damage. But in some cases, a supervolcano erupts with catastrophic results. One of the most famous supervolcanoes is Krakatoa on an island between Sumatra and Java, in Indonesia. Krakatoa had been dormant for centuries but then came 'back to life' in 1883. On 26 August 1883, two thirds of the island was blown away. Ash rose 27 km into the atmosphere and the explosions could be heard 4600 km away in Australia!

What happens when plates move towards each other and collide?

The Earth's crust that lies under the continents is called **continental crust**. The crust that lies under the oceans is called **oceanic crust**. The continental crust is usually thicker and contains less dense rocks, such as granite. The oceanic crust contains denser rocks like basalt.

When a continental plate and an oceanic plate collide, the denser oceanic plate sinks below the lighter continental plate. This process is called **subduction**. Part of the oceanic crust may be forced into the mantle where it forms liquid magma. At the same time, the continental crust gets squashed and pushed into **folds**.

Over millions of years, this results in the formation of mountain ranges. This is what happened and continues to happen in South America as the Nazca Plate collides with the American Plate to create the Andes Mountains (Figure 8.14).

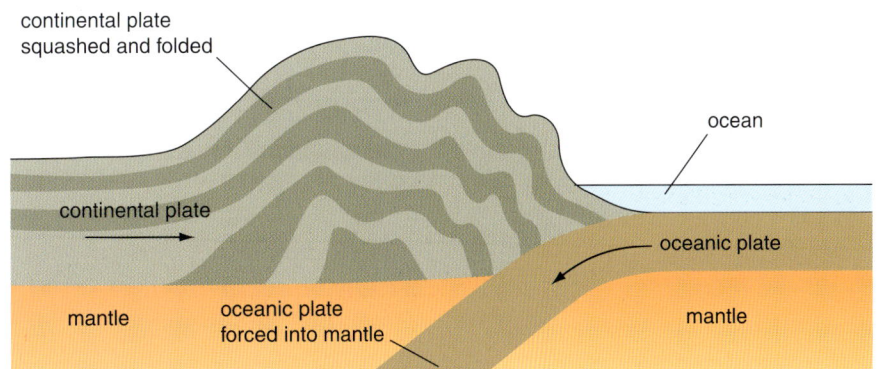

Figure 8.13 When an earthquake occurs below the seabed, it can set off a giant wave called a tsunami. This is what happened in the Indian Ocean on 26 December 2004, causing the loss of more than 200 000 lives and enormous damage in the areas affected.

continental plate
squashed and folded

ocean

continental plate

oceanic plate

mantle

oceanic plate
forced into mantle

mantle

Figure 8.14 A continental plate and an oceanic plate moving towards each other and colliding

If two continental plates or two oceanic plates move towards each other, they tend to collide head on. Layers in the plates become tilted, folded and even turned upside down. This process has been happening for millions of years during the formation of the Alps, the Pyrenees and the Himalayas.

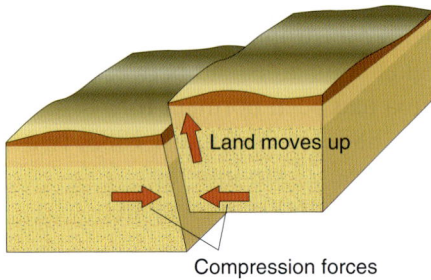

Figure 8.15 When plates collide and push against each other, the Earth may crack forming a reverse fault

Sometimes cracks, rather than folds, appear when plates collide and push against each other with huge compression forces. Land on one side of the crack is forced up above the other side which may then get buried (Figure 8.15). This is called a **reverse fault**.

❽ The photo in Figure 8.16 is taken looking east.
  a) Which directions did the forces come from to create the fold?
  b) Was the force involved a compression or a tension force?
  c) Draw a sketch of the fold and explain how it formed.

Figure 8.16 Folds in the Earth's crust at Stair Hole, Lulworth Cove, in Dorset. Folds like this have been created by the movement of the Eurasian Plate.

Breaking news – Massive Earthquake

We report the latest situation
BBC News 24
  For further information go to:
  www.geo.ed.ac.uk and search on 'quakes'
  or www.eqnet.org

Activity – Earthquake

Use the internet websites given above to find out about a recent large earthquake.
Now imagine you are the senior reporter for a news programme on TV.
You are about to report live about the earthquake. Compose what you will say.
Your report must not last more than 2 minutes.

Important things that you should report on are:
- the time and place of the earthquake;
- injuries to people, the number of people missing and any loss of life;
- damage to homes and other buildings in the area;
- the size of the quake on the Richter Scale.

Other things that you may wish to include are:
- any indications that the earthquake may recur with aftershocks;
- whether the earthquake might have been predicted;
- the dates of any previous earthquakes in the area;
- one or two personal experiences of people affected by the earthquake.

8.6 Can we predict when earthquakes and volcanic eruptions will occur?

Most earthquakes occur in predictable areas of the world along or near the boundaries between tectonic plates. One of the areas most at risk from earthquakes is along the San Andreas fault in California, where the Pacific plate slides past the North American plate.

Although we can predict *where* earthquakes are likely to occur, we cannot predict *when* they will occur with any accuracy. If we could predict when an earthquake will occur, even to the nearest day or so, people could be evacuated from their homes to safety and many lives could be saved.

Predicting earthquakes

Figure 8.17 A seismograph used to monitor for the risk of earthquakes in the Philippines

Earthquakes occur suddenly and without warning. Predicting them is difficult but there are signs to look for and monitoring instruments that scientists can use.
- **Seismometers** are used continuously in risky earthquake areas to monitor for any possible smaller shocks before the main earthquake occurs.
- **The GPS (global positioning system)**, which locates positions on the Earth to within a few centimetres, can be used to follow the movement of plates.
- **Strainmeters** are used to measure the forces in rocks.
- **Tiltmeters** are used to record tiny movements and bulges in rocks.
- **Animals often behave strangely** before an earthquake. Snakes come out of hibernation, rats leave their holes and cattle become restless.

In some cases, scientists have been remarkably successful in predicting earthquakes. For example, in 1975 Chinese scientists in Haicheng City recorded increased seismic activity and had reports that snakes were coming out of hibernation. People living in the city were warned and told to leave their homes. At 7.30p.m. an earthquake occurred, destroying most of the buildings in Haicheng, but thousands of people's lives were saved.

Predicting volcanic eruptions

Figure 8.18 After being quiet for nearly 400 years, the volcano on the Caribbean island of Montserrat erupted in the summer of 1997. The capital, Plymouth, had to be abandoned because of the flows of hot rocks, ash and gases.

It is much easier to predict when a volcano will erupt than when an earthquake will occur. Unlike earthquakes, volcanoes usually give a variety of warning signs months, sometimes years, before they erupt. Here are some of the warning signs that scientists pick up prior to a volcanic eruption.
- **Seismometers** will detect any small earth tremors caused by the movement of magma deep inside a volcano.
- **The GPS** will detect any changes in ground level around a volcano as pressure builds up inside it.

- **Air monitoring equipment** can be used to monitor the levels of sulfur dioxide near the mouth of a volcano. An increase in the concentration of sulfur dioxide indicates that the magma is rising.
- **Infra-red cameras** mounted on satellites will detect any increase in ground temperature near a volcano as magma rises towards the Earth's surface.

Although these instruments can give us clues that a volcano might soon erupt, just as with earthquakes, predicting an eruption is only an intelligent guess.

In 1985, the ground rose by 2 metres in just a few days at Potsuoli near Naples in Italy, close to the volcano on Mount Vesuvius. Buildings were in danger and 40 000 people were evacuated from the area. Scientists are still waiting for the eruption!

8.7 How has the Earth's atmosphere changed?

Using radioactive dating techniques scientists estimate that the oldest rocks on the Earth were formed about 4500 million (4.5 billion) years ago. This is usually taken to be the age of the Earth.

- During the first billion (1000 million) years of the Earth's existence, there was intense volcanic activity. Rocks decomposed, elements reacted and gases were released to form the first atmosphere.

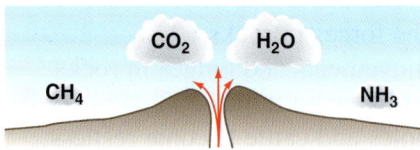

Figure 8.19 The early atmosphere on Earth was mainly carbon dioxide and water vapour.

This early atmosphere was mainly carbon dioxide and water vapour with smaller proportions of methane (CH_4) and ammonia (NH_3) (Figure 8.19).

- As the molten rocks on the Earth's surface cooled and the temperature dropped further, most of the water vapour condensed to form rivers, lakes and oceans (Figure 8.20).
- Plants evolved and first appeared on the Earth 3500 million years ago. As plants slowly colonised most of the Earth's surface, further changes occurred in the atmosphere. Plants took in water and carbon dioxide for photosynthesis and released oxygen (Figure 8.21).

Figure 8.20 As the Earth cooled down, most of the water vapour in the early atmosphere condensed to form rivers, lakes and oceans.

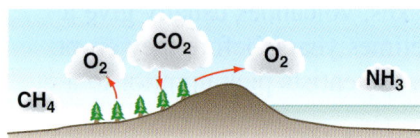

Figure 8.21 When plants appeared on the Earth, carbon dioxide and water were taken up during photosynthesis and oxygen was produced.

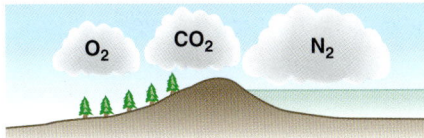

Figure 8.22 Methane and ammonia burnt in the oxygen from photosynthesis producing more water, more carbon dioxide and nitrogen.

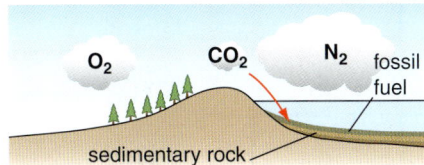

Figure 8.23 Over billions of years, carbon dioxide became locked up as fossil fuels and in sedimentary rocks as carbonates.

- As oxygen collected in the atmosphere, flammable gases, like methane and ammonia, burnt in this oxygen producing more water, more carbon dioxide and nitrogen (Figure 8.22).
- At the same time, carbon dioxide in the changing atmosphere was being removed by two other processes:
  1 the formation of fossil fuels from carbon compounds in plants and sea creatures;
  2 the deposition of carbonates as sedimentary rocks, following erosion by rivers and from the shells and bones of sea creatures.

So, over billions of years, most of the carbon dioxide in the air gradually became locked up as fossil fuels and in sedimentary rocks as carbonates (Figure 8.23).

⑮ Write a word equation for the reaction which occurred when methane reacted with oxygen in the Earth's early atmosphere.

⑯ In the early atmosphere, ammonia reacted with oxygen to form nitrogen and water vapour. Copy and balance the following equation for this reaction.

$$....NH_3 + ...O_2 \rightarrow ....N_2 + ...H_2O$$

⑰ What two substances do the Earth's surface and atmosphere contain which are essential for all living things?

⑱ How did these two substances get into the atmosphere and onto the Earth's surface?

8.8 Our atmosphere today

Our atmosphere has remained more or less the same for the last 200 million years. It is composed of:
- about four fifths (80%) nitrogen;
- about one fifth (20%) oxygen;
- small proportions of other gases including carbon dioxide, water vapour and noble gases.

Accurate percentages of these gases in dry air are shown in Table 8.1.

The Earth is the only planet in our Solar System with oxygen in its atmosphere and abundant surface water in rivers, lakes and oceans. Other planets, such as Mars, do however have some water vapour and polar ice caps.

The noble gases

The noble gases are in Group O on the extreme right of the Periodic Table (Figure 8.24).

Gas	Percentage
Nitrogen	78.1
Oxygen	20.9
Argon	0.9
Carbon dioxide	less
Neon	than
Krypton	0.1
Xenon	

Table 8.1 The percentages of gases in dry air

	3	4	5	6	7	Group O
						He
	B	C	N	O	F	Ne
		Si	P	S	Cl	Ar
					Br	Kr
					I	Xe
						Rn

Transition metals

Poor metals

Halogens Noble gases

Figure 8.24 The position of the noble gases in the Periodic Table

The noble gases are all colourless and odourless with very low melting points and boiling points. They all exist as separate single atoms. Other gaseous elements (hydrogen H_2, oxygen O_2, nitrogen N_2 and the halogens) all exist as diatomic molecules.

Until 1962, there were no known compounds of the noble gases. Chemists thought they were completely unreactive. Because of this, they were called the *inert* gases. Today, several of their compounds are known and have been produced. They are not inert, so we now call them the *noble* gases. The word 'noble' was chosen because unreactive metals like gold and silver are called *noble metals*.

Uses of the noble gases

Helium is used in balloons and airships because it has a low density and is non-flammable.

The noble gases produce a coloured glow when their atoms are bombarded by a stream of electrons. The stream of electrons can be produced either from a high-voltage discharge across the terminals of a discharge tube or from a laser. Neon and argon are used in discharge tubes to create fluorescent advertising signs. Neon tubes give a red colour and argon tubes give a blue colour.

Argon and krypton are used in electric filament lamps (light bulbs). If there is a vacuum inside the lamps, metal atoms evaporate from the superhot tungsten filament. To reduce this evaporation and prolong the life of the filament, the bulb is filled with an unreactive gas which cannot react with the hot tungsten filament.

Figure 8.25 A technician releases a weather balloon filled with helium

19 Hydrogen was once used for inflating balloons.
a) What was the special advantage of using hydrogen?
b) Why was the use of hydrogen dangerous?

20 a) Give two reasons why argon and krypton are used in light bulbs (filament lamps).
b) Why is air unsuitable to use inside filament lamps?

Figure 8.26 The bright lights of Piccadilly

Activity – What changes are occurring in the Earth's atmosphere?

Table 8.2 shows the concentration of carbon dioxide in the atmosphere, the average global temperature and the global population at intervals of 50 years since 1750.

	Year					
	1750	1800	1850	1900	1950	2000
Concentration of carbon dioxide in the atmosphere / % by volume	0.0278	0.0282	0.0288	0.0297	0.0310	0.0368
Average global temperature / °C	13.3	13.4	13.4	13.6	13.8	14.4
Global population / millions	350	500	1000	1500	3000	5500

Table 8.2

1 It is widely believed that the burning of fossil fuels has increased the level of carbon dioxide in the atmosphere.
a) Natural gas is probably the simplest fossil fuel containing mainly methane with traces of ethane. Copy and complete the following equation for the burning of methane.

$$CH_4 + \ldots\ldots O_2 \rightarrow CO_2 + \ldots\ldots$$

b) State two major processes which involve burning fossil fuels.
c) Name one important process that removes carbon dioxide from the atmosphere.

❷ Use the data in Table 8.2 to plot a graph which will show conclusively that the level of carbon dioxide in the atmosphere is increasing.
   a) In your graph, what was i) the independent variable; ii) the dependent variable?
   b) From your graph, describe the way in which carbon dioxide is increasing.

❸ Many people believe that the increasing levels of carbon dioxide in the atmosphere are responsible for global warming.
   a) Why should we worry about global warming?
   b) How does carbon dioxide contribute to global warming? (You may wish to refer to Section 6.4 to help you with this question.)
   c) Which rows of data in Table 8.2 show that there might well be a link between carbon dioxide and global warming?
   d) Name one other gas that contributes to global warming.

❹ Some people believe that global warming is largely the result of an increasing global population using ever-increasing amounts of fossil fuels. What are your views about this suggestion?

❺ What should we in the UK be doing about global warming:
   a) at a personal / family level;
   b) at a national / government level?

Summary

✓ The Earth and its atmosphere provide the raw materials for everything we need.

✓ The Earth is nearly spherical with a layered structure comprising:
   ● a thin **crust**;
   ● a **mantle** extending almost halfway to the Earth's centre, which has all the properties of a solid except that it can flow very slowly;
   ● a central **core** of about half the Earth's diameter, made of iron and nickel, the outer part of which is liquid and the inner part of which is solid.

✓ The Earth's crust and the upper part of the mantle are cracked into a number of pieces called **tectonic plates**.

✓ The decay of natural radioactive materials inside the Earth releases heat which produces convection currents in the mantle. These convection currents cause the tectonic plates to move at relative speeds of a few centimetres per year.

✓ This movement of tectonic plates (called **continental drift**) is normally very slow, but on occasions it can be sudden and disastrous. At the boundaries of the plates, it can result in **earthquakes** and **volcanic eruptions**.

✓ For the last 200 million years, proportions of the different gases in the Earth's atmosphere have been more or less the same:
   ● about four fifths (80%) nitrogen;
   ● about one fifth (20%) oxygen;
   ● small proportions of other gases including carbon dioxide, water vapour and the noble gases.

✓ The **noble gases** occupy Group O of the Periodic Table. They are chemically unreactive. Helium is much less dense than air and is used in balloons. Neon and argon are used in fluorescent advertising signs and argon and krypton are used in electric filament lamps (light bulbs).

✓ During the first billion years of the Earth's existence, there was intense volcanic activity.

- This activity released gases that formed the early atmosphere – mainly carbon dioxide and water vapour with smaller proportions of methane and ammonia.
- As the Earth cooled down, the water vapour condensed to form rivers, lakes and oceans.
- When plants evolved, carbon dioxide and water were used up in photosynthesis and oxygen was produced. Although the reverse happened when plants respired, the proportion of oxygen in the atmosphere slowly increased.

✓ During the next 3 billion years, most of the carbon from the carbon dioxide in the air gradually became locked up in fossil fuels and in sedimentary rocks as carbonates.

✓ During the last 150 years or so, the level of carbon dioxide in the atmosphere has slowly increased. This is due to the burning of fossil fuels in industry (particularly power stations), in our homes and in our vehicles.

✓ The increase in the level of carbon dioxide in the atmosphere has resulted in **global warming**.

EXAMQUESTIONS

1 a) Figure 8.27 shows the layered structure of the Earth. Copy and complete the figure by adding the *three* missing labels.

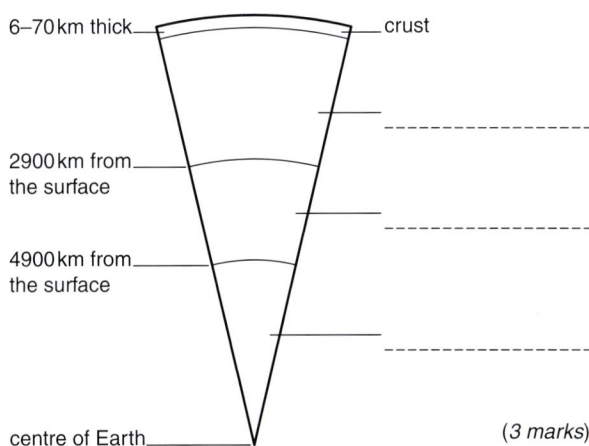

Figure 8.27

6–70 km thick — crust

2900 km from the surface

4900 km from the surface

centre of Earth

(3 marks)

b) The crust and upper mantle are cracked into a number of pieces called tectonic plates.
   i) Why do these tectonic plates move?
   (2 marks)
   ii) Explain how the movement of the tectonic plates can lead to earthquakes.
   (4 marks)

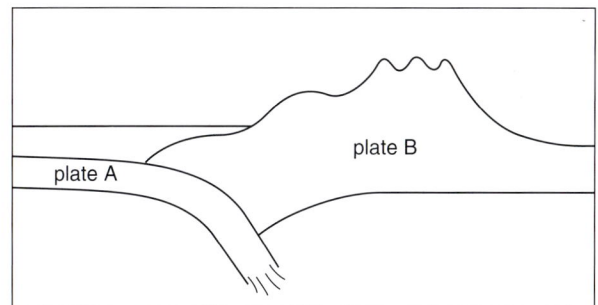

plate B

plate A

Figure 8.28

2
a) Figure 8.28 shows a cross section through two tectonic plates A and B. Copy the diagram and add the following labels: ocean; oceanic crust; continental crust; folds; mountains; sediments; mantle.
(7 marks)
b) Put an arrow on your diagram to show the movement of plate B.
(1 mark)
c) In one year, will plate A move a few millimetres, a few centimetres or a few metres?
(1 mark)
d) Why does plate A sink below plate B?
(1 mark)
e) What is the scientific name for this process?
(1 mark)
f) Mark a point, X, on your diagram where solid rock may be forming a viscous liquid.
(1 mark)

❸ Match the terms A, B, C and D with the spaces 1–4 in the sentences.
A continental drift
B land mass
C radioactive processes
D tectonic plates   (*4 marks*)

In 1915, Alfred Wegener put forward the idea that millions of years ago there was a single large ..... (1) ..... This broke up and the smaller parts, which we now call ..... (2) ....., moved apart. This process is called ..... (3) ..... and the heat required for the movement comes from ..... (4) ..... in the Earth's core and mantle.

❹ a) Name the gases labelled A–E in the pie charts in Figure 8.29.   (*5 marks*)

The Earth's early atmosphere

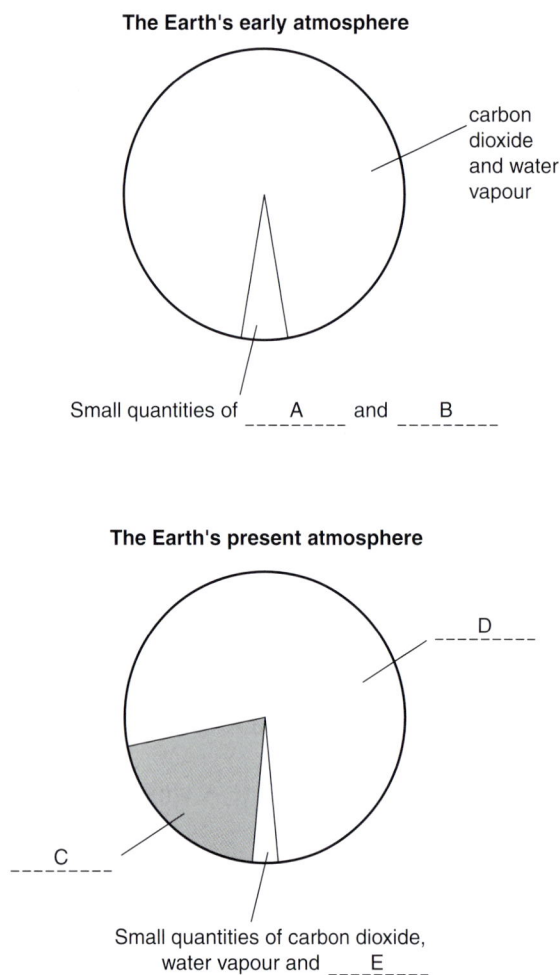

carbon dioxide and water vapour

Small quantities of _____ A _____ and _____ B _____

The Earth's present atmosphere

D

C

Small quantities of carbon dioxide, water vapour and _____ E _____

Figure 8.29

b) Over the last 150 years, the amount of carbon dioxide in the atmosphere has increased from 0.03% to 0.04%.
   i)   Why is the amount of carbon dioxide gradually increasing?   (*2 marks*)
   ii) Suggest two effects that increasing levels of carbon dioxide may have on the environment.   (*2 marks*)
c) The percentage of carbon dioxide in country air is often lower than that in city air. Why is this?   (*1 mark*)

❺ The gases in our present atmosphere evolved over billions of years from the gases produced by erupting volcanoes.
a) The gas from erupting volcanoes contained a high proportion of water vapour, but our present atmosphere contains only a very small proportion of water vapour. Why is this?   (*4 marks*)
b) Volcanic gases also contained a high proportion of carbon dioxide. Suggest three processes which led to a reduction of carbon dioxide in the atmosphere.   (*3 marks*)
c) Oxygen did not appear in the Earth's atmosphere for about 1 billion (1 000 000 000) years. Why was this?   (*2 marks*)
d) Scientists believe that nitrogen was formed in our atmosphere when ammonia (NH_3) from the volcanic gases reacted with oxygen. Copy and complete the following equation for this reaction.

$$...NH_3 + ...O_2 \rightarrow ...N_2 + ...H_2O$$
        (*2 marks*)

Chapter 9
How is heat transferred and what is meant by energy efficiency?

At the end of this chapter you should:

✓ understand how heat (thermal energy) is transferred;

✓ know the factors that affect the rate at which heat is transferred;

✓ appreciate ways of reducing the transfer of heat into and out of bodies;

✓ understand how energy can be transformed (changed) from one form into another;

✓ appreciate that a device is more efficient if a greater percentage of the energy supplied is usefully transformed.

Figure 9.1 Using energy efficiently benefits the planet and saves us money

How is energy transferred by radiation?

Radiation is the transfer of thermal energy (heat) from one place to another by means of electromagnetic waves.

Energy is transferred from the Sun to the Earth by electromagnetic waves. (See Section 11.2.) These waves, which include visible light waves and invisible **infra-red** waves, can travel through a vacuum.

Although infra-red waves cannot be seen you can feel them. When infra-red waves are absorbed they cause a heating effect. This is why you feel warm in the sun. So, infra-red waves are often called **thermal radiation** or just **radiation** for short.

All bodies transfer energy by thermal radiation. The hotter a body is the more energy it transfers each second by thermal radiation. A warm object will radiate only infra-red, but at higher temperatures it may also emit visible light waves. At about 800 °C, an object may glow 'red hot'. In this state it is emitting infra-red waves as well as waves from the red end of the visible light spectrum.

All bodies emit and absorb thermal radiation. But which surfaces are better emitters and which are better absorbers?

Emitting radiation

Dark-coloured matt surfaces emit radiation at a faster rate than light-coloured shiny surfaces (Figure 9.3). (A matt surface is dull and non-shiny.)

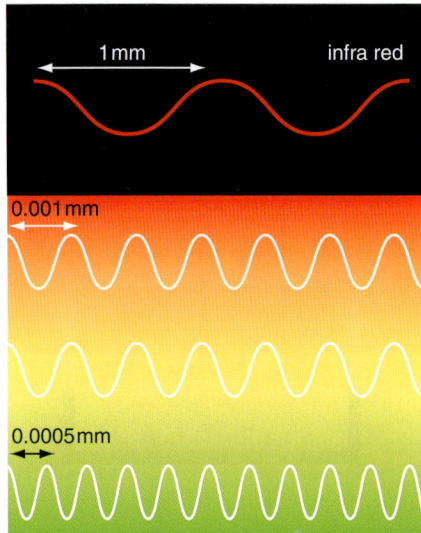

not drawn to scale

Figure 9.2 Very hot objects emit infra-red and visible light waves

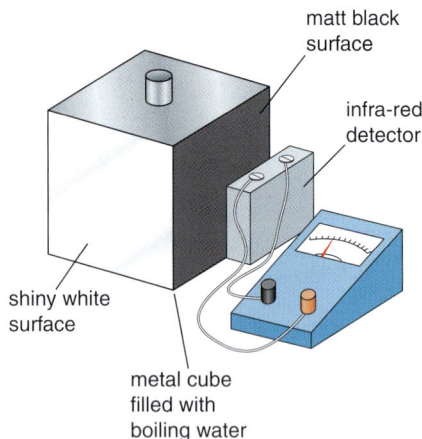

Figure 9.3 Each side of the cube has a different colour or texture but all the sides are at the same temperature. The infra-red detector shows the relative amount of infra-red radiation emitted by each surface. It shows the highest reading when opposite the matt black side of the cube, and the lowest reading when opposite the shiny white side of the cube.

Variables

A variable is something that is subject to variation such as the different surfaces of the cube in Figure 9.3.

There are different kinds of variables:

Continuous variables can have any numerical value such as the length of a piece of string or the mass of a stone.

Discrete variables are restricted to whole numbers, for example, the number of layers of insulation or the number of atoms in a molecule.

Ordered variables have a clear order of size or mass or length, such as small, medium and large pairs of socks.

Categoric variables are different types of something such as the different surfaces of the cube in Figure 9.3 or the different sexes.

Later in this section and in Section 5.7, we will meet two other types of variable – independent variables and dependent variables.

The infra-red radiation emitted by a body can also be detected using an infra-red camera. The camera produces an image that shows different temperatures as different colours. (See also Section 11.3.)

Absorbing radiation

Objects heat up when they absorb thermal radiation more quickly than they emit it.

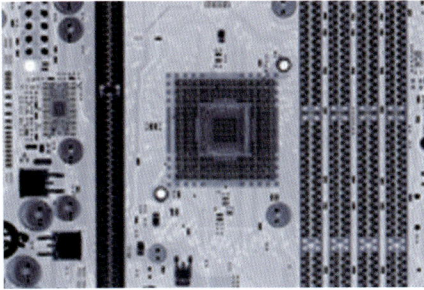

Figure 9.4 Electronic components are sometimes joined to a 'heat sink'. The heat sink has black metal fins which radiate energy away from the component, helping to keep it cool.

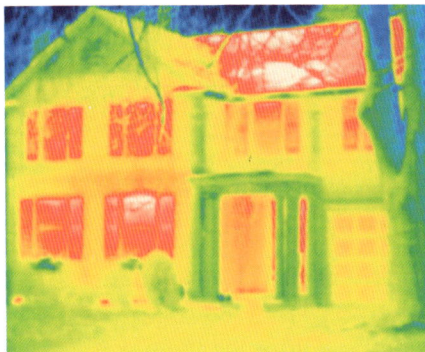

Figure 9.5 An infra-red photograph can show which parts of a building lose the most heat. The owner may be able to add insulation or take measures to reduce heat loss.

Figure 9.6 The outside metal of a dark-coloured car warms up quicker in the Sun than the outside metal of a light-coloured car. A polished shiny car will not get as hot as an unpolished dull car.

The example given in Figure 9.6 shows that:
- dark-coloured surfaces absorb thermal radiation faster than light-coloured surfaces;
- shiny surfaces reflect more thermal radiation than dull, matt surfaces.

So, good emitters of thermal radiation are also good absorbers of thermal radiation.

---

❶ The window blinds in Ms Clymo's laboratory are matt black. The blinds are often closed. On a sunny day, Ms Clymo records the laboratory temperature every 15 minutes between 9.00a.m. and 11.15a.m. Her results are shown in Figure 9.7.
   a) At what time did Ms Clymo close the blinds in her laboratory? Explain fully the reason for your answer.
   b) At 11.00a.m. the laboratory is absorbing 10 kJ of energy every second from the Sun. At what rate must the laboratory be losing energy? Give a reason for your answer.
   c) Although Ms Clymo used a thermometer to measure the temperature, she could have used a temperature sensor and data logger. What are the advantages of using a temperature sensor and data logger rather than a thermometer?
   d) Estimate the probable temperature of the

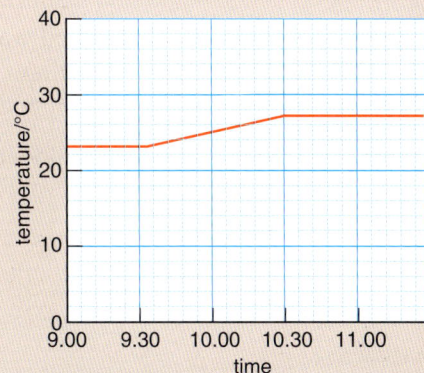

Figure 9.7 A graph showing the temperature inside Ms Clymo's laboratory

laboratory at 3.00p.m. Explain why your estimate is unlikely to be reliable.

❷ What happens to the temperature of a body that emits more thermal radiation than it absorbs?

Activity – Does the intensity of infra-red radiation increase as you move towards the source of radiation?

To answer this question, Ted designed a simple experiment. He used a thermometer to detect infra-red radiation. When the intensity of the radiation absorbed by the thermometer went up, the reading on the thermometer also went up.

The apparatus Ted used is shown in Figure 9.8.

Ted wrote this plan for his experiment.
- Place a thermometer with a blackened bulb 60 cm in front of the infra-red lamp and record the reading.
- Switch on the lamp.
- Wait two minutes then take the thermometer reading again.
- Move the thermometer 10 cm at a time towards the lamp.
- Each time the thermometer is moved wait two minutes and then take the new reading.
- Repeat the experiment.

The thermometer readings after the lamp was switched on are recorded in Table 9.1.

Distance from lamp	Thermometer reading		
	1st	2nd	mean
60	37	38	37.5
50	41	40	40.5
40	44	44	44.0
30	51	50	50.5
20	73	73	73.0

Table 9.1 The results of Ted's experiment

The temperature recorded before the lamp was switched on was 22 °C.

❶ Why did Ted take the temperature before the lamp was switched on?

❷ Why did Ted use a thermometer with a blackened bulb, rather than a clear bulb?

❸ Why did Ted wait two minutes before taking each temperature?

❹ Give two reasons why the second set of temperatures was not exactly the same as the first set.

❺ Ted moved the thermometer 10 cm between temperature readings. This is called the

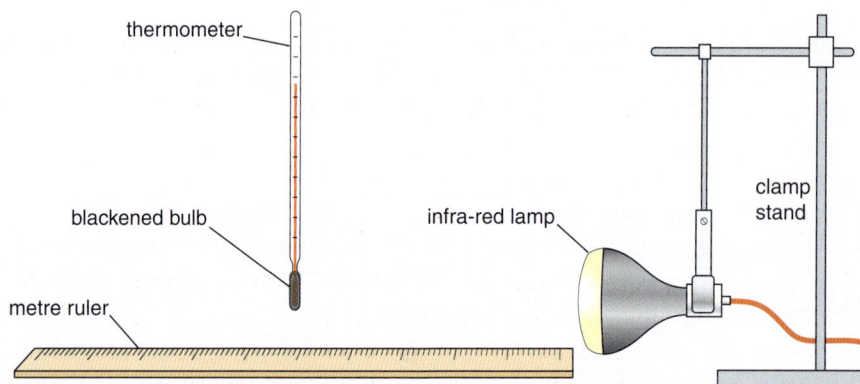

Figure 9.8 Apparatus used by Ted to investigate infra-red radiation

interval. What advantage would Ted gain by reducing the interval to 5 cm?

6 A results table should always include both the quantity measured and the unit in which the quantity is measured. What is missing from Ted's results table?

7 In an experiment, the **independent variable** is the quantity that you change. The **dependent variable** is the quantity you measure when the independent variable changes.

a) Is the independent variable in this experiment distance or temperature?

b) What is the dependent variable in this experiment?

c) Is temperature in this experiment a categoric variable, a continuous variable or an ordered variable?

8 Ted repeated the experiment in order to improve the **reliability** of the results. Reliable results are results that you can trust and that can be reproduced. Are Ted's results reliable? Give a reason for your answer.

9 The **sensitivity** of a measuring instrument refers to the smallest change that the instrument can detect. A thermometer that can be read to the nearest 0.1 °C is more sensitive than one that can be read to the nearest 1 °C. Was the thermometer Ted used sensitive enough for this experiment? Give a reason for your answer.

10 Drawing a graph or putting results in a table often helps us to identify **anomalous results**. An anomalous result is one that does not fit the expected pattern. Are any of Ted's results anomalous?

11 The way in which we present results depends on the type of variable they represent. If the independent variable is an ordered variable or a categoric variable, the results are best presented as a bar chart. If the independent variable is a continuous variable, the results are best presented as a line graph. How would the results of the cube experiment in Figure 9.3 be best presented?

12 A set of results has a certain range. The **range of results** is from the smallest to the largest value.

a) What is the range of Ted's results?

b) How could Ted have increased the range of his results?

13 If the results or the evidence from an experiment are to be taken seriously, they must be **valid**. Valid results are reliable and they answer the question asked. Explain why the results and evidence from Ted's experiment are valid.

9.2

How is energy transferred by conduction and convection?

Radiation is the transfer of heat (thermal energy) by waves. Radiation does not involve particles of matter. Unlike radiation, the transfer of heat by conduction and convection does involve the movement of particles.

Conduction

Conduction is the transfer of heat (thermal energy) between materials in contact, or between different parts of the same substance, without the materials or the substance moving.

If you walk around barefoot, you will soon notice that ceramic floor tiles feel much colder to your feet than carpet. Your feet feel cold because they are losing energy to the tiles in the form of heat. This transfer of heat from your feet to the tiles is an example of **conduction**.

All metals are very good conductors, but materials like plastic, wood and glass are poor conductors. Gases are very poor conductors. These poor conductors are called **insulators**.

If one end of a metal bar is placed in a Bunsen flame, the heat from the flame is quickly transferred along the bar from the hot end to the cold end. This happens because metals contain both atoms and mobile electrons (electrons that are free to move).

At the heated end of the metal bar (Figure 9.9a), the energy from the flame:

- increases the kinetic energy of the atoms, making them vibrate faster and with a bigger amplitude. These atoms collide with their neighbours passing on energy so they also vibrate faster. This process transfers energy slowly through the bar;
- increases the kinetic energy of the mobile electrons. The rapid movement of the mobile electrons in the hotter parts of the metal is transferred via collisions to adjacent electrons. These, in turn, transfer energy to other electrons and energy is rapidly conducted through the metal.

So, energy is transferred along a metal bar by a series of collisions between neighbouring atoms and by the movement of mobile electrons.

a) In a metal energy is conducted quickly by fast-moving, mobile electrons.

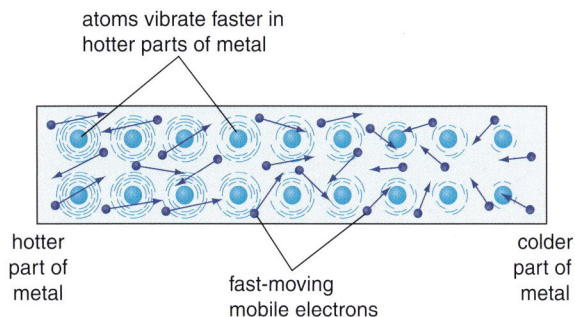

b) In an insulator, there are no mobile electrons, so energy is conducted slowly by atoms colliding as they vibrate.

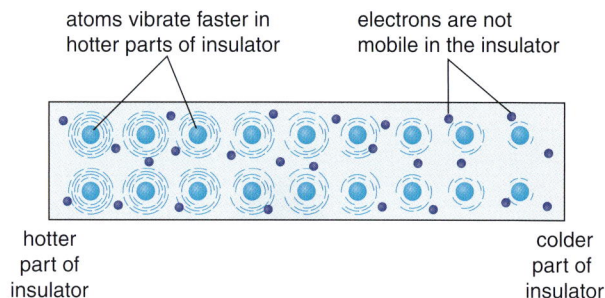

atoms vibrate faster in hotter parts of metal

hotter part of metal

fast-moving mobile electrons

colder part of metal

atoms vibrate faster in hotter parts of insulator

electrons are not mobile in the insulator

hotter part of insulator

colder part of insulator

Figure 9.9 Conduction in metals and insulators

In thermal insulators there are no mobile electrons. So, the transfer of thermal energy relies on the collisions between neighbouring atoms (Figure 9.9b). This is usually a very slow process.

Convection

Convection is the transfer of heat (thermal energy) by the movement of a liquid or gas due to differences in density.

Convection is a second way in which thermal energy (heat) can be transferred from one place to another. Convection occurs in both liquids and gases (fluids). During convection, the warmer liquid or gas moves, transferring its extra energy with it.

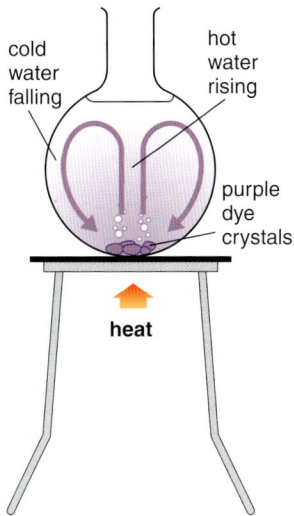

Figure 9.10 Energy transfer in a heated liquid

Figure 9.10 shows what happens when water is heated.

As the water at the bottom of the flask is heated, the water molecules gain energy. This extra energy causes:

- the molecules to move around each other faster;
- the molecules to move apart and take up more space;
- the water in the warmer region to expand;
- the warm water, as it expands, to become less dense than the cooler water around it;
- the less dense warm water to rise.

As the warm water rises, cooler water flows in to replace it. In due course, this water also gets heated and rises. These movements of hot and cold water, due to changes in density, are called **convection currents**. Convection currents stop when all parts of the water are at the same temperature.

❹ A metal jug and a plastic container are taken out of a refrigerator. They are both at the same temperature. Why does the metal jug feel colder?

❺ Explain, in terms of particles, how heat is transferred from a gas flame through the bottom of an aluminium saucepan to a liquid in the pan.

❻ Why are non-metals better insulators than metals?

Figure 9.11 Which parts of the mountaineer are insulated best from the cold? Which parts of his body will lose the most heat?

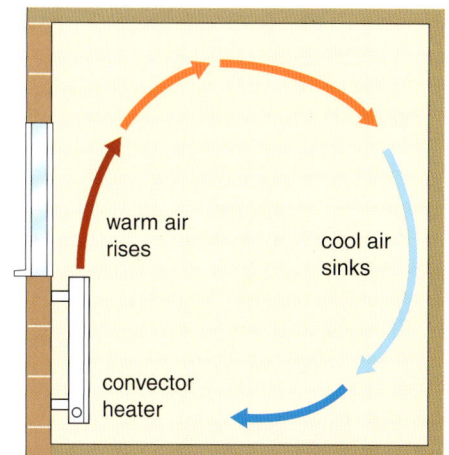

Figure 9.12 A convector heater creates an air flow in a room. Energy is transferred through a gas in the same way as a liquid.

Coping with the cold

Animals and humans try to cope with the cold in various ways.

A seal has a thick layer of fat all around its body. The fat is a good insulator, so it reduces the rate of energy transfer from the seal.

Animal fur traps pockets of air, which is a good insulator. On a cold day an animal will fluff up its fur to trap more air. In this way, it can stay warm even in very cold weather.

Humans wear clothes to keep warm. When it is very cold, it is wise to put on several layers of clothes. Air is then trapped between the layers as well as in the fabric of the clothes.

Figure 9.13 Trapped air in its fur keeps the rabbit warm

7 Barbara, who runs Hot Snak, a takeaway food shop, is fed up with customers complaining that the coffee they buy goes cold too quickly. She decides to carry out a simple experiment to find a disposable cup which will keep coffee hot for at least 10 minutes. Barbara fills four cups, made from two different materials, with the same volume of boiling water. She puts lids on *two* of the cups then waits for 10 minutes before taking the new water temperature.

Cup	Material	Lid	Temperature after 10 minutes / °C
A	Cardboard	Yes	60
B	Cardboard	No	52
C	Polystyrene	Yes	75
D	Polystyrene	No	66

Table 9.2

Table 9.2 shows the results of Barbara's experiment.

a) What did Barbara do to make this experiment a **fair test**? A fair test is one in which only the independent variable affects the dependent variable.
b) Barbara only recorded one set of results. Why would it have been better if she had repeated the experiment and obtained two more sets of results?
c) Which results should Barbara compare before deciding to use cardboard or polystyrene cups?
d) Which results should Barbara compare before deciding if it is worth providing a lid for the cups?
e) What type of variables are cardboard and polystyrene in Barbara's experiment?

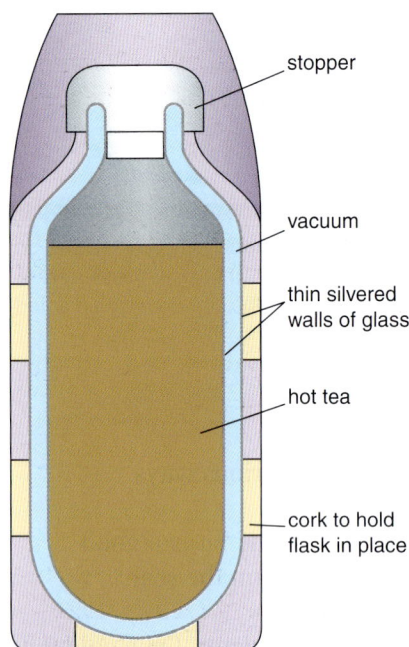

Figure 9.14 A vacuum flask reduces energy transfer

Labels: stopper; vacuum; thin silvered walls of glass; hot tea; cork to hold flask in place

How does a vacuum flask keep hot drinks hot or cold drinks cold?

A vacuum flask keeps drinks hot (or cold) by reducing energy transfer by conduction, convection and radiation.

A vacuum flask reduces energy transfer by having:
- a vacuum between the double walls of the container to reduce energy transfer by conduction and convection;
- walls with shiny surfaces to reduce energy transfer by thermal radiation;
- a stopper made of a good insulator such as cork or plastic, to reduce energy transfer by conduction and convection.

9.3 Do all bodies (objects) transfer heat at the same rate?

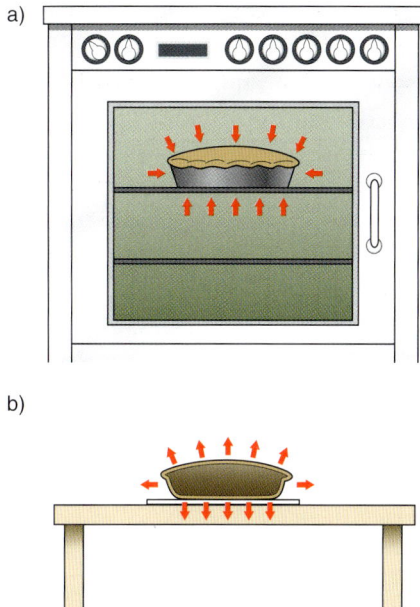

a)

b)

Figure 9.15 a) Heat is transferred to the pie from the hot air inside the oven. b) Heat is transferred from the hot pie to the surroundings.

The words 'body' and 'bodies' are general terms used to refer to *any* object or objects. So, although a glass beaker, a metal spoon, and a wooden toy are different objects each one can be referred to as a 'body'.

Heat (thermal energy) is transferred from a body to its surroundings when the body is at a higher temperature than the surroundings.

The rate at which a body transfers heat depends on various factors:
- the type of material the body is made from;
- the shape of the body;
- the dimensions (size) of the body;
- the difference in temperature between the body and its surroundings;
- what the body is in contact with.

Type of material
Bodies made from different materials will, under similar conditions, transfer heat at different rates. So, two hot bodies made from different materials may start at the same temperature, but after a few minutes their temperatures will be different. The two bodies will transfer heat and cool down at different rates.

Shape
The rate of heat loss from a body depends on its surface area. Changing the surface area by changing the shape of the body will change the rate of heat transfer (Figure 9.16).

Dimensions (size)
The rate of heat loss from a body depends on the surface area of the body. But the rate at which the temperature of a hot body falls (or a cold body rises) depends on its surface area to volume ratio.

In general, for bodies of the same shape, the rate of temperature change increases as the size decreases. This means that smaller bodies cool down faster than larger ones.

Temperature of the surroundings
The bigger the temperature difference between a body and its surroundings, the faster the rate at which heat is transferred. The graph in Figure 9.17 shows how the heat lost from a hot water pipe increases as the temperature difference between the pipe and the surrounding air increases.

What the body is in contact with
When a body is in contact with a conductor it will lose heat faster than when it is in contact with an insulator. In a building, heat loss is reduced by using insulating materials. (See Section 9.4.)

Figure 9.16 Cooling fins help to lower the temperature of the motorbike engine. They do this by greatly increasing the surface area of the engine allowing more cooling air to come into contact with more hot metal.

How can the rate of heat transfer be reduced?

Keeping your home warm

Keeping your home warm is not just about turning on the central heating or lighting a fire. It's also about reducing the amount of heat transferred from inside your home to the air outside.

The features which allow animals to keep warm or lose heat are studied in Section 3.1.

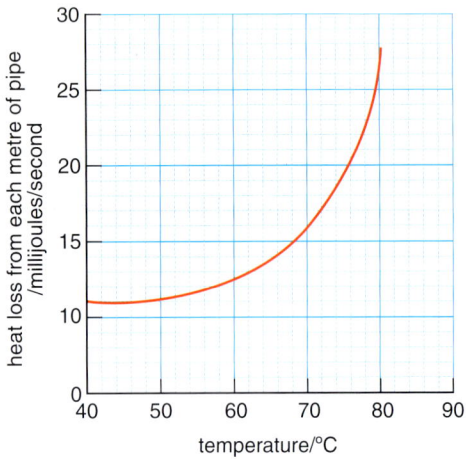

Figure 9.17 Heat loss from a hot water pipe

8 Figure 9.18 shows a section of a car radiator. Explain how the design of the radiator helps to cool the car engine.

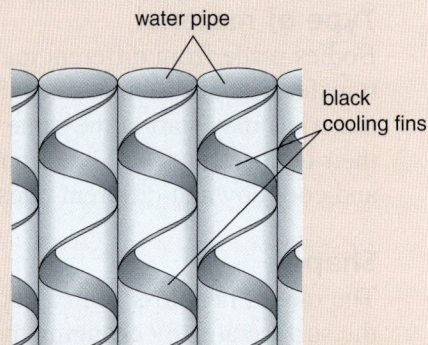

Figure 9.18

9 Figure 9.19 shows three ball bearings in a beaker of hot water. The ball bearings are left to reach the same temperature as the water. They are then taken out and placed on a table.
   a) Which ball bearing cools down at the slowest rate? Explain the reason for your choice.
   b) Explain how the rate of cooling of the ball bearings would change if instead of being placed on a table they had been put in a refrigerator.
   c) Explain why the different sizes of the ball bearings can be regarded as an ordered variable.

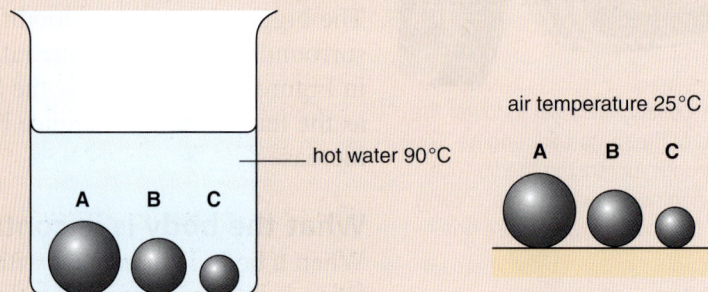

Figure 9.19

Different amounts of heat are transferred through the roof, the walls, the windows, the floor and the doors of a house. These are shown in Figure 9.20.

Reducing heat transfer involves trying to stop conduction and convection. Figure 9.21 shows some of the methods used to reduce the heat loss from our homes.

Most methods of reducing heat transfer work by trapping air. If air is trapped in small pockets it cannot move far, so heat loss by convection is greatly reduced. This trapped air is also a good insulator so heat loss by conduction is reduced (Figure 9.22).

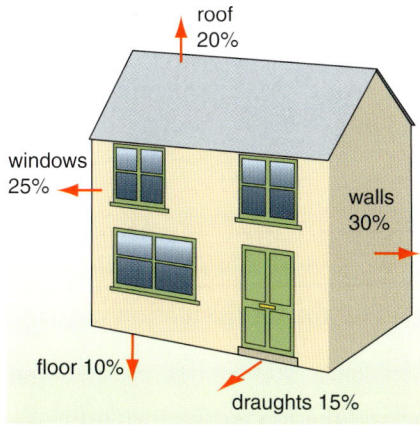

Figure 9.20 Heat loss from an average house

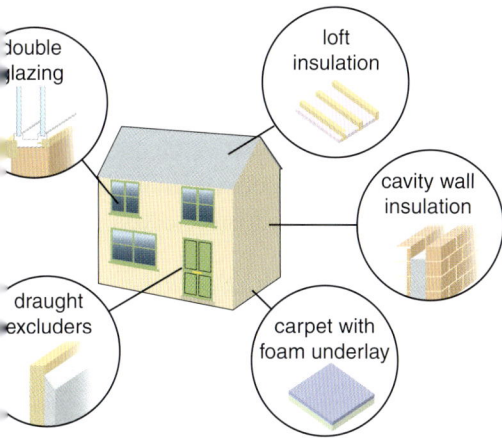

Figure 9.22 Air trapped by fibre wool and by foam

With double glazing both the air and the glass are good insulators so very little heat is lost by conduction. To keep heat loss by convection low, the layer of air between the glass sheets must be thin.

Draught excluders trap warm air inside the house and stop cold air coming in.

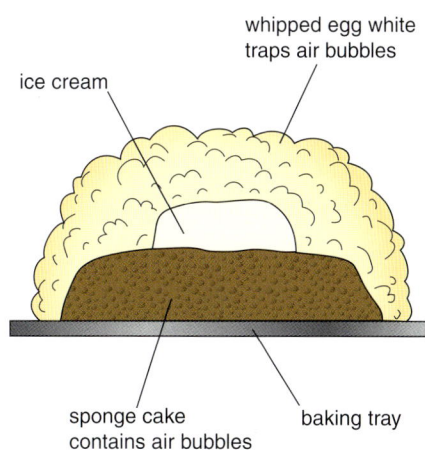

Figure 9.21 Reducing heat loss from our homes

⑩ Give three ways of reducing heat transfer from a house.

⑪ Explain why fish and chips wrapped in layers of paper stay hot for a long time.

⑫ Figure 9.23 shows a pudding called 'baked Alaska'. The pudding is baked in a very hot oven for a few minutes. Why does the ice-cream not melt?

Saving energy, saving money

Installing any type of insulation costs money. However, once installed less energy will be needed to keep a home warm. This means that heating bills will be less, so you save money.

The time it takes to get back the cost of the installation from the money saved on the heating bills is sometimes called the 'pay-back time'. It may take 40 years for the money saved by having double-glazed windows to pay for their installation.

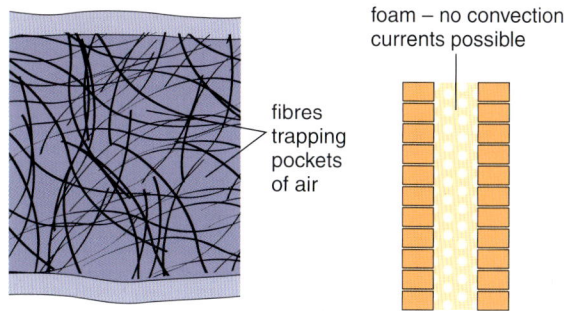

Figure 9.23 A baked Alaska

Example

Gurpal lives in a 1930s house which was built with very little insulation. Gurpal has just installed loft insulation. It cost £180. The insulation will save Gurpal £45 each year on his heating bill. What is the 'pay-back time' for the insulation?

$$\text{pay-back time} = \frac{\text{cost of insulation}}{\text{money saved each year}} = \frac{180}{45} = 4 \text{ years}$$

So, Gurpal will get back the cost of installing the loft insulation in four years.

Generally, the shorter the pay-back time and the longer the insulation lasts, the more cost-effective it is. So, cost-effective insulation over its lifetime, will save far more money on energy bills than the initial purchase and installation cost.

Consider Gurpal's loft insulation. After four years the insulation has been paid for from energy savings. But those savings continue until the insulation needs replacing. Assuming the insulation lasts 30 years and energy costs stay the same, the total amount saved on energy bills is £45 × 30 = £1350. All for an initial outlay of £180. This is cost-effective!

⑬ Table 9.3 gives information about different types of insulation.
   a) Calculate the reduction in the rate of energy loss for each type of insulation.
   b) Which type of insulation gives the largest energy saving for each pound (£) spent?

⑭ Table 9.4 gives the costs, savings and replacement times for two methods of reducing energy loss in the home.

Which method is the most cost-effective for reducing energy loss over 20 years?

Type of insulation	Installation cost in £s	Energy loss before installation in J/s	% Reduction in energy loss after installation
Cavity wall	450	1500	30
Loft	250	1000	40
Carpet	800	400	25

Table 9.3

Method of reducing energy loss	Cost to install (£)	Yearly saving (£)	Replacement time (years)
Draughtproofing	75	25	5
Temperature controls on radiators	120	20	20

Table 9.4

More energy saving ideas

Close the curtains

Thick curtains stop cold air blowing into a room. They also trap air between the window and the curtain so that less heat is lost by conduction. Closing curtains costs nothing; what could be more cost effective than this?

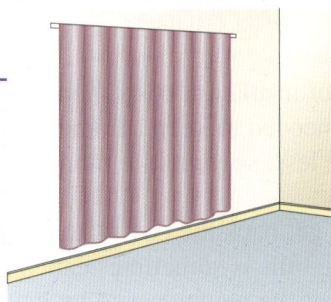

Figure 9.24

Insulate the hot water tank

Fit a thick jacket around the hot water tank. Fibres in the jacket trap small pockets of air so that less heat is lost by conduction and convection.

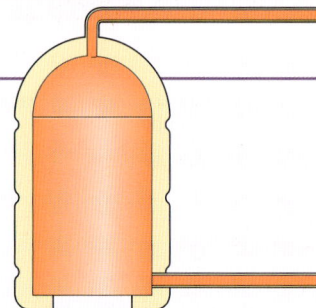

Figure 9.25

Career – building services engineer

A building services engineer is involved in the design and installation of water, heating, lighting, electricity and ventilation in a building. In fact a building engineer is involved in everything needed to make a home, school or workplace safe, healthy and comfortable. Building services engineers try to make buildings as energy efficient as possible by reducing waste energy to a minimum. In this way, they help to reduce both the use of our energy resources and any environmental pollution from the building. A building services engineer needs to understand the science of energy transfer so that they can advise builders and householders on the best ways to make a house energy efficient.

9.5 Energy efficiency

Energy from nothing!

We often hear that the Earth's non-renewable energy resources are running out. At the same time, we are encouraged to use more renewable energy resources and to develop these resources. But, wouldn't it be great if we could create energy from nothing? Unfortunately this is impossible!

Energy can be transferred (moved) from one place to another or transformed (changed) from one form into another, but it cannot be created or destroyed. This is a fundamental principle called the **law of conservation of energy**.

Energy is a bit like money. Money can make things happen when it is transferred from one person to another. In a similar way, energy can make things happen, but only when it is transferred.

Figure 9.26 Chemical energy in the muscles of the sprinter is transformed into kinetic energy, heat in her muscles and a small amount of sound energy.

Figure 9.27 A firework transforms chemical energy to light, heat, sound and gravitational potential energy.

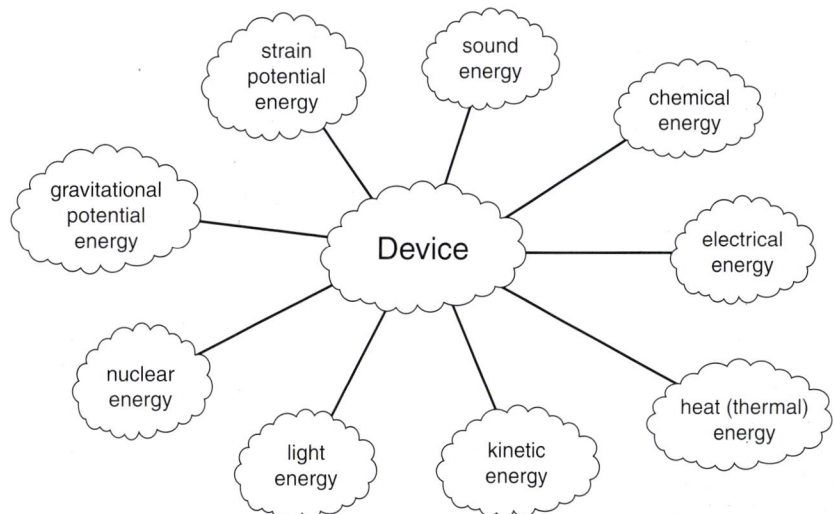

Figure 9.28 A device does not consume (use up) energy. It transforms energy from one form to another.

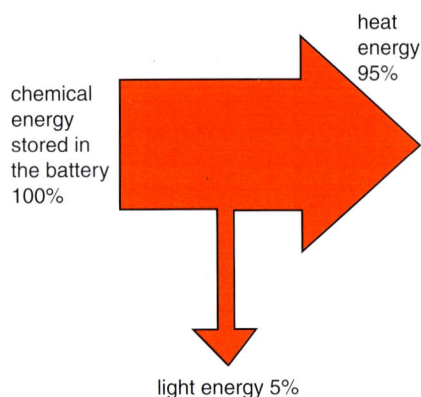

Figure 9.29 A Sankey diagram for a torch

heat energy 95%

chemical energy stored in the battery 100%

light energy 5%

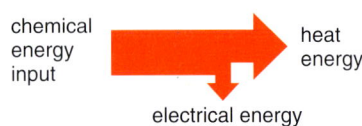

Figure 9.30 A Sankey diagram for a coal-burning power station

chemical energy input

heat energy

electrical energy

A **Sankey diagram** can be used to show what happens to the energy transformed by a device or a machine. The wider the arrow in the diagram the greater the energy transformed.

Figure 9.29 shows that most of the chemical energy stored in the torch battery is transformed *not* into light but into heat. Only the light energy is wanted – the heat energy is wasted, but the total amount of energy stays the same. The energy has been conserved.

Any device that transfers or transforms energy also wastes some of the energy. The device transfers only part of the energy to where it is wanted and in the form that is wanted. This is the useful energy. The rest of the energy is transformed into forms that are not wanted. This is not useful and the energy is wasted.

A coal-burning power station is designed to transform chemical energy from coal into electrical energy. But it also transfers a lot of energy into heat, which is wasted.

In every energy transfer or transformation some energy, however small, is transferred to the surroundings. This causes the surroundings to become warmer. As a result, molecules in the air gain kinetic energy and move a little faster. As the energy spreads out, any increase in temperature is usually too small to notice. The energy has been transferred to millions of molecules which all move a tiny bit faster.

As the energy spreads out, more molecules share the energy and it becomes more difficult to use it for further energy transformations. So the energy that started in a useful form is wasted, *but it has not vanished*.

In any energy transfer or transformation, some energy spreads out and becomes less useful to us.

15 A ball does not bounce back to the same height from which it is dropped. Explain how the law of conservation of energy applies to a bouncing ball.

16 Whenever Steve and Sue go out for the evening, Sue always switches the lights and TV off. Sue says that switching things off when you don't need them saves energy. But Steve is puzzled. Steve thinks that because energy cannot be destroyed it must always be there, so it cannot be saved. Explain carefully, using the idea of energy conservation, what Sue means by 'saving energy'.

How good is a machine at transferring useful energy?

Efficiency is the proportion of the energy input or energy supplied that is transferred into a useful form by a device.

The **efficiency** of a machine or device tells us how good it is at transferring energy into a useful form or forms. A device would be 100% efficient if the total energy going in were the same as the energy transferred in a useful form.

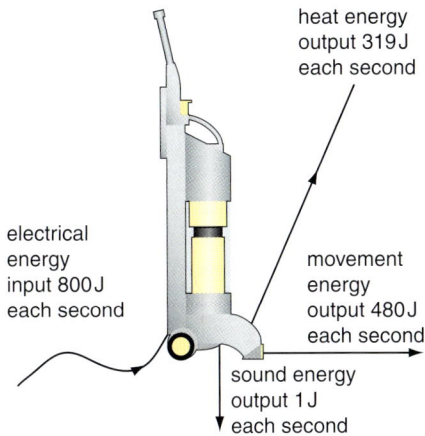

heat energy
output 319 J
each second

electrical
energy
input 800 J
each second

movement
energy
output 480 J
each second

sound energy
output 1 J
each second

Figure 9.31 The energy transfers produced by a vacuum cleaner

The efficiency of a device can be calculated using the following equation.

$$\text{efficiency} = \frac{\text{useful energy transferred by the device}}{\text{total energy supplied to the device}}$$

Example

A vacuum cleaner is designed to transfer electrical energy into kinetic energy. But it also transfers energy as heat and sound. What is the efficiency of the vacuum cleaner in Figure 9.31?

$$\text{efficiency} = \frac{\text{useful energy transferred by the device}}{\text{total energy supplied to the device}}$$

$$\text{efficiency} = \frac{480}{800} = 0.6$$

The efficiency is 0.6. Sometimes, efficiency is given as a percentage. So in this case it would be:

$$\frac{480}{800} \times 100 = 60\%$$

The greater the proportion, or percentage, of the energy that is usefully transferred by a device, the more efficient it is.

17 An electric oven is described as being 70% efficient. What does this mean?

18 A diesel engine transforms 40% of the input energy to kinetic energy, 15% to sound and 45% to heat.
  a) Draw a Sankey diagram for the diesel engine.
  b) What is the efficiency of the engine?

Low-energy light bulbs

A 100 W filament lamp is only about 5% efficient. Most of the electrical energy is transformed into heat. This energy is wasted.

Low-energy light bulbs are much more efficient at transforming electrical energy to light energy. Because of this a 20 W low-energy bulb can give out as much light as a 100 W filament lamp (Section 11.3).

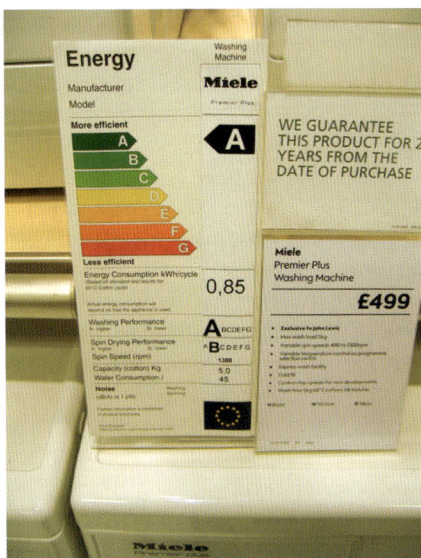

a)

b)

Figure 9.33 a) A filament light bulb   b) A low-energy light bulb

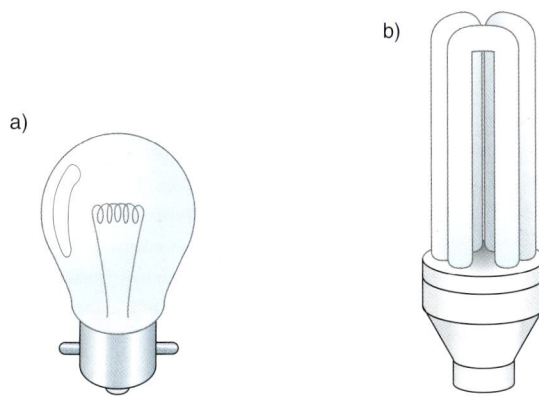

Figure 9.32 This European Energy Label must be displayed on all new fridge freezers (as well as other types of electrical appliances). The information on the label allows you to compare the efficiency and running costs of different models.

Activity – Investigating the efficiency of an electric motor

When an electric motor is used to lift a weight it transforms electrical energy into useful gravitational potential energy. Susan decided to investigate how the efficiency of this transfer depends on the weight being lifted. She used the apparatus shown in Figure 9.34.

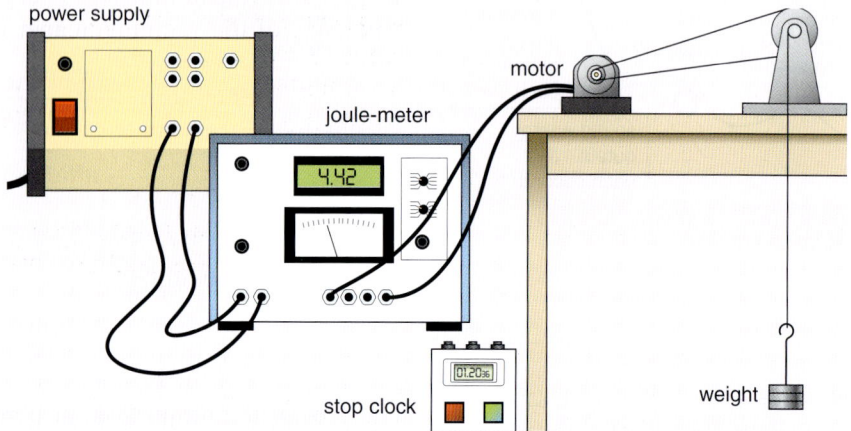

Figure 9.34 The apparatus used to investigate the efficiency of an electric motor

Susan used a joulemeter to measure the electrical energy supplied to the motor each time it lifted a weight 0.82 m. She then repeated the experiment, using a range of different weights. For each weight, she obtained two measurements and recorded the average. Her results are shown in Table 9.5.

Electrical energy input in joules	Weight lifted in newtons	Calculated efficiency as a percentage
15	2	10.7
17	3	14.2
21	4	15.3
26	5	15.5
33	6	14.6
46	7	12.2

Table 9.5 The results of Susan's experiment

❶ In Susan's investigation:
   a) what range of weights was used?
   b) what was the independent variable?
   c) what was the dependent variable?
   d) which variable was controlled during the investigation?

Remember, the independent variable is the quantity you change. The dependent variable is the quantity that changes because of this.

❷ Why did Susan take two measurements for each weight?
❸ Draw a graph of weight lifted (horizontally) against percentage efficiency (vertically).
❹ How does the efficiency of this motor depend on the weight being lifted?

How can 'waste' energy be useful?

In some devices and machines heat that would normally be transferred to the air, and wasted, can be usefully used. For example, some of the waste heat from a car engine can be used to warm the air inside the car.

The efficiency of a coal-burning power station is about 35%. Most of the energy from the fuel is wasted as heat. Although a combined heat and power (CHP) station is less efficient at electricity generation it uses some of the 'waste' energy to heat water. Pipes carry the hot water to local buildings where it is used for heating. A CHP scheme is worthwhile where there is a steady demand for heating.

Laundries waste a lot of hot dirty water. Figure 9.35 shows how the heat from the dirty water can be used to warm clean water. This reduces the energy consumption and the cost of running the laundry.

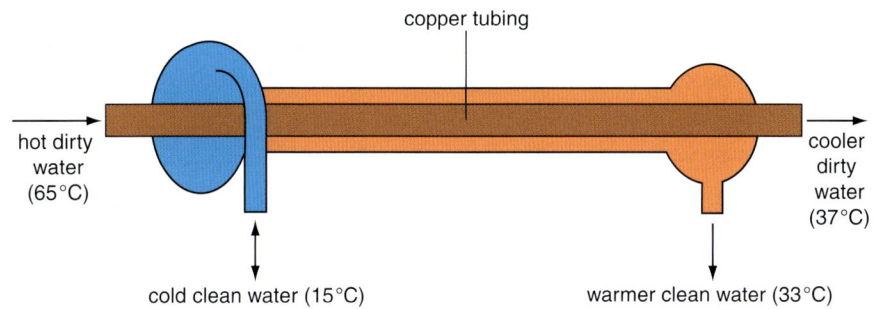

Figure 9.35 Using a heat exchanger reduces running costs

Even the gas given off from sewage can be used. The gas is collected, burnt and the energy released is used to generate electricity.

⑲ The Sankey diagram in Figure 9.36 shows that a street lamp transforms electrical energy to light and heat. What is the efficiency of the street lamp?

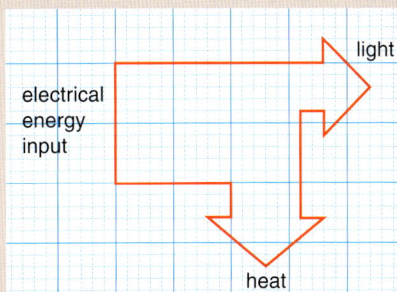

Figure 9.36

⑳ Erika has a very old but working freezer. Explain why it is probably more cost-effective to replace the freezer with a new 'A' energy rated freezer.

㉑ In a traditional power station 65% of the energy input is transferred and wasted as heat. Figure 9.37 shows the energy transfers in a combined heat and power (CHP) station.

Why is the CHP station more efficient than a traditional power station?

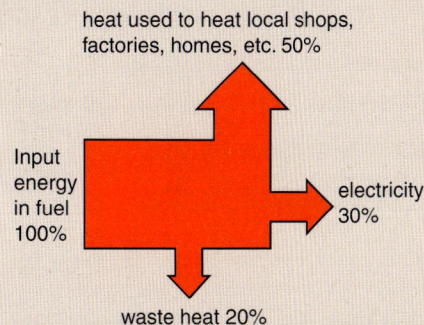

Figure 9.37

Summary

✓ **Infra-red waves (thermal radiation)** can travel through a vacuum.

✓ All bodies emit thermal radiation.

✓ Dark-coloured matt surfaces emit and absorb thermal radiation better than light-coloured shiny surfaces.

✓ Shiny surfaces reflect thermal radiation better than dull, matt surfaces.

✓ The transfer of heat by **radiation** does not involve the movement of particles, unlike the transfer of heat by **conduction** and **convection**.

✓ A material that contains free electrons is a good **conductor** of heat.

✓ **Convection currents** in a fluid occur because of changes to the density of the fluid.

✓ The rate at which a body transfers heat depends on:
  • the type of material the body is made from;

  • the shape of the body;
  • the dimensions of the body;
  • the difference in temperature between the body and its surroundings.

✓ **The law of conservation of energy** says: 'energy cannot be created or destroyed. It can only be transformed from one form to another form.'

✓ In any energy transfer or transformation, the energy spreads out and becomes less useful to us. Some energy is always wasted during the transfer or transformation.

✓ **Efficiency** is the proportion of the energy input or energy supplied that is transferred in a useful form by a machine or device.

✓ $$\text{Efficiency} = \frac{\text{useful energy transferred by the device}}{\text{total energy supplied to the device}}$$

EXAMQUESTIONS

❶ Figure 9.38 shows a cross section through the outside wall of a house.

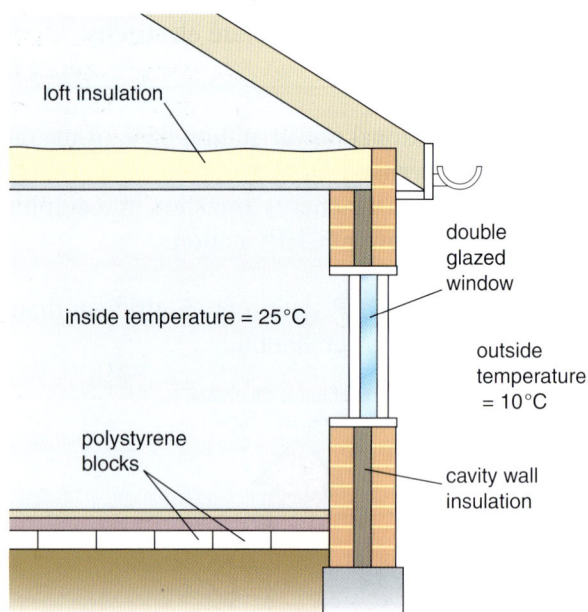

Figure 9.38

loft insulation

inside temperature = 25°C

polystyrene blocks

double glazed window

outside temperature = 10°C

cavity wall insulation

a) Explain how the polystyrene blocks under the floor reduce heat loss from the house. *(2 marks)*

b) What happens to the rate of heat loss through the window when the outside temperature drops to 5 °C? Assume the temperature inside the house stays at 25 °C. Give a reason for your answer. *(2 marks)*

Type of insulation	Heat loss before insulation (J/s)	Heat loss after insulation (J/s)
Floor	700	420
Loft	1600	940
Cavity wall	2200	900
Double glazing	1140	780

Table 9.6

(c) Table 9.6 gives information about four different ways to insulate a house. Which type of insulation in Table 9.6 is most effective in reducing heat loss from the house? To gain full marks you must support your answer with calculations. *(2 marks)*

2 Marion and Jim have been asked to find out which of the three materials K, L or M would be the best for making a winter coat. They each produce a plan for their investigation.

Marion's plan
i   From each material, cut a rectangle 8 cm by 20 cm.
ii Wrap material K round an empty metal can; hold it in place with an elastic band.
iii Pour 200 cm³ of boiling water into the can; place a thermometer in the water.
iv Wait until the temperature reaches 85 °C then start a stop watch.
v   Take the temperature every minute for 10 minutes.
vi At the end of 10 minutes throw the water away.
vii Repeat the experiment with material L and then with material M.
viii Each time use 200 m³ of water and wait until the temperature reaches 85 °C before starting the stop watch.

Jim's plan
i   Cut a piece of material from each of K, L and M.
ii Wrap that material from K around an empty metal can.
iii Pour in some hot water.
iv Start a stop watch and take the temperature every few minutes.
v   When the water has cooled down throw it away.
vi Start again with the other two materials.
    a) Give four reasons why Marion's results will be more valid than Jim's results.   (*4 marks*)

b) Marion's results are presented in Figure 9.39.
    Which material should the winter coat be made from? Give a reason for your answer.   (*2 marks*)

3 The Sankey diagram in Figure 9.40 shows what happens to the input energy for a television.
a) Calculate the efficiency of the TV. Show clearly how you work out your answer.   (*2 marks*)

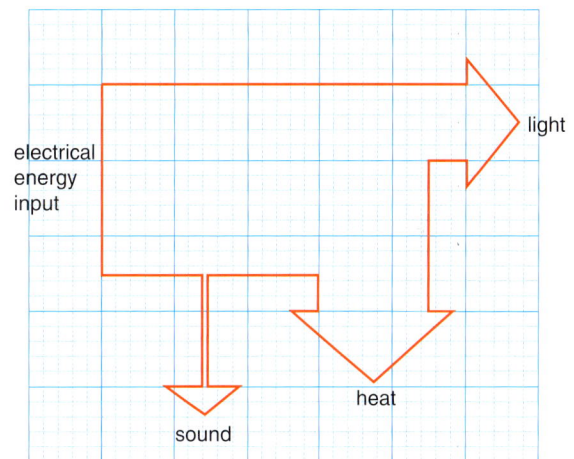

Figure 9.40

b) Joe is arguing with a friend about the conservation of energy. Joe says that since energy is conserved, it must always be there, so there is no point in switching things like a TV or computer off. Explain why Joe is wrong.   (*2 marks*)

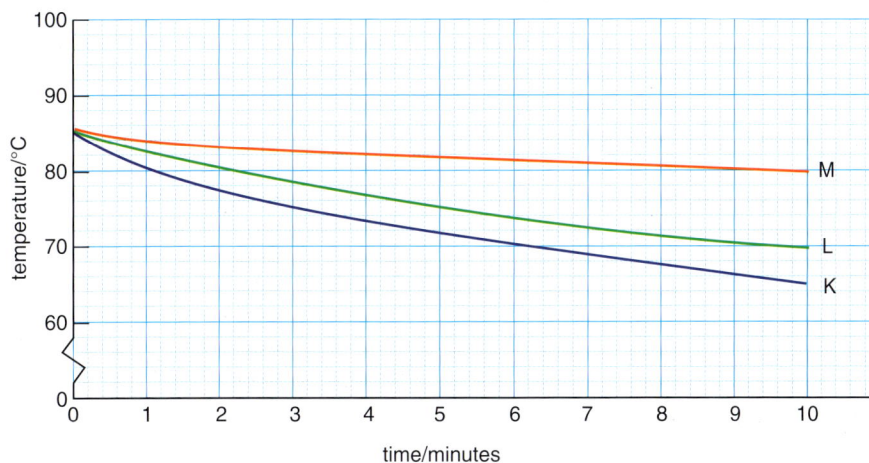

Figure 9.39 A graph showing Marion's results

Chapter 10
How do we generate and use electricity?

At the end of this chapter you should:

✓ know that electrical devices transform electrical energy to other forms of energy at the flick of a switch;

✓ know that the power of a device is the rate at which it transforms energy;

✓ understand how the National Grid transfers electricity from power stations to consumers;

✓ know how electricity is generated in power stations;

✓ know that electricity is generated using a variety of energy sources;

✓ appreciate that each type of energy source used to generate electricity has its advantages and disadvantages;

✓ be able to calculate the amount of electrical energy transferred from the mains supply and the cost of this energy.

Figure 10.1 Different energy sources are used to generate electricity. In the UK most electricity is still generated in power stations using the energy trapped in fossil fuels

Why is electrical energy so useful?

In most homes there are various devices (appliances) which work by transforming electrical energy to other forms of energy.

In industry, electrical energy is a widely used source of energy.

Electrical energy is such a useful form of energy because it is easily transformed into other forms of energy, such as:
- heat (thermal energy);
- light;
- sound;
- kinetic energy;
- gravitational potential energy.

This is why we have so many devices that are designed to work from the electricity mains supply.

Figure 10.3 shows five everyday electrical devices and the energy transformations they are designed to bring about.

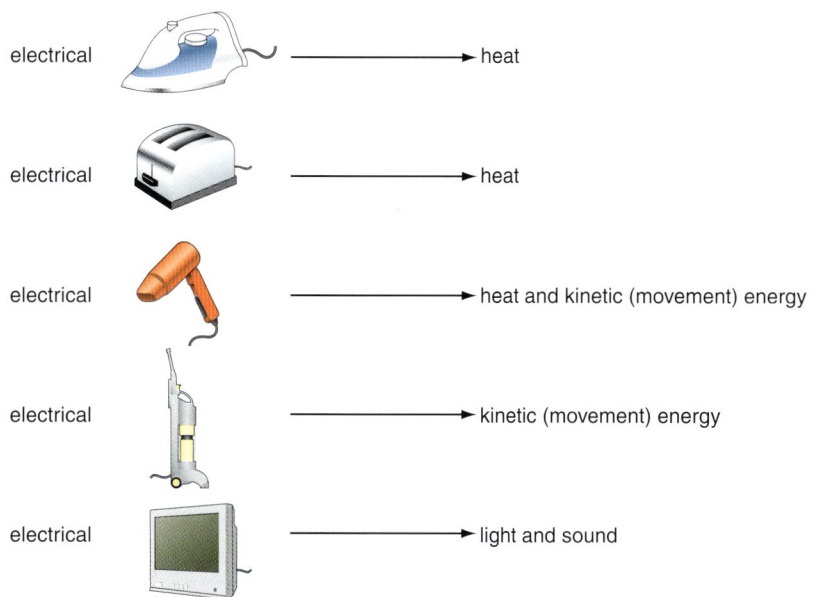

Figure 10.2 Many machines work by transforming electrical energy

Figure 10.3 Some everyday electrical devices designed to transform energy

❶ Describe the energy transformations that each of the following electrical devices are designed to bring about.

ipod   mobile phone charger
drill   washing machine

❷ Name three devices designed to transform electrical energy into heat energy.

Electrical energy and power

The amount of electrical energy that a device transforms depends on two things:
- how long the device is used;
- the rate at which the device transforms energy (uses electricity).

The rate at which a device transforms energy is called its **power**.

Power is measured in watts (W) or joules per second (J/s).

Power is the rate at which a device transforms energy.

Figure 10.4 An electric jigsaw

Figure 10.5 The information plate shows the power of the jigsaw

A device that transforms 1 joule of energy every second, from one form to another, has a power of 1 watt. So a 350 watt electric jigsaw (Figure 10.4) will transform 350 joules of electrical energy to other forms of energy every second it is switched on.

$$1 \text{ watt} = 1 \text{ joule/second } (1 \text{ W} = 1 \text{ J/s})$$

Power can also be measured in kilowatts (kW).

$$1 \text{ kilowatt (kW)} = 1000 \text{ watts (W)}$$

Table 10.1 shows the power of some of the devices mentioned earlier in this section.

Appliance	Power rating
Lamp	60 W
Television	150 W
Toaster	250 W
Vacuum cleaner	770 W
Hairdryer	1200 W
Iron	1800 W

Table 10.1 The power rating of some everyday electrical devices

3 Each of the following appliances is used for 15 minutes.

1200 W hairdryer   1.8 kW iron
100 W light bulb   20 W radio

a) Which appliance transfers the most energy?
b) Which appliance transfers the least energy?

Give a reason for each of your choices.

Activity – Using electrical devices at home

1 Look around your own home. How many different devices can you name that work from the mains electricity supply? All you have to do is plug in and flick the switch.

2 Figure 10.6 shows two types of fan. Type A is a 230 V, 50 W mains operated fan. Type B is a 3 V, low-power battery operated fan.

Compare the advantages and disadvantages of using each type of fan.

Figure 10.6 Type A – mains operated fan   Type B – battery operated fan

Paying for electricity

Energy is transferred from the mains electricity supply every time an appliance is plugged in and switched on. This energy must be paid for.

Appliances with a high power rating (see Section 10.1) transfer a lot of energy from the mains electricity supply each second. For example, a 1.8 kW iron transfers 1800 joules of electrical energy every second. So, any appliance with a high power rating costs a lot to use.

The cost also depends on how long the appliance is switched on for. The longer the appliance is on, the more it costs.

If the total amount of energy transferred to a home was measured in joules, the numbers would be enormous. So, electricity companies measure the electrical energy transferred using a much larger unit. The unit they use is the **kilowatt-hour** (kWh).

The energy transferred by an appliance can be calculated using the equation:

energy transferred = power × time
(in kilowatt-hours)   (in kilowatts) (in hours)

Example
A 4 kilowatt electric cooker is switched on for 3 hours. How much electrical energy is transferred from the mains supply to the cooker?

energy transferred = power × time
= 4 × 3 (power in kilowatts and time in hours)
= 12 kilowatt-hours (kWh)

Calculating the bill

The cost of using mains electricity can be calculated by simply multiplying the total number of kilowatt-hours of energy transferred by the cost of one kilowatt-hour.

total cost = number of kilowatt-hours × cost per kilowatt-hour

Example
Between February and May a homeowner uses 247 kilowatt-hours of electrical energy. Assuming one kilowatt-hour costs 12p, what is the cost of using this energy?

total cost = number of kilowatt-hours × cost per kilowatt-hour

= 247 × 12p = 2964p = £29.64

---

4. Estimate the amount of energy you would save each year by replacing just one 100 W light bulb with a 20 W energy-efficient bulb.

5. Now estimate the amount of energy you would save each year if you replaced all the light bulbs in your home in this way. (Assume that all the light bulbs are 100 W.)

6. Finally, estimate how much energy we could save each year if everyone in the UK did this. (Assume there are 20 million homes in the UK, all with three bedrooms.)

7. Calculate the cost of using each of the following appliances. Assume 1 kWh of electrical energy costs 12p.
   a) A 2 kW heater switched on for 4 hours.
   b) A 9 kW shower used for 10 minutes.
   c) A 2500 W kettle switched on for 2 minutes.
   d) A 60 W light bulb switched on for 45 minutes.

Activity – Working out the cost of using electricity

Find the information plate on an electrical appliance that you have at home. The information plate will be on the outside of the appliance. There is no need to take anything to pieces!

❶ What is the power of the appliance? Write the power in watts (joules per second) and in kilowatts.

❷ Estimate how long the appliance is used in one week. Write the time used in seconds and in hours.

❸ Calculate, in joules and in kilowatt-hours, how much electrical energy is transferred to the appliance in one week.

❹ Calculate the weekly cost of using the appliance. (Assume 1 kWh costs 12p.)

❺ Suggest a way to make your use of the appliance more energy efficient. For example, if the appliance is a dishwasher, you should only use it fully loaded.

10.3 The National Grid

A network of cables and transformers links our homes, offices, schools and factories to the power stations that generate electrical energy. This network is called the **National Grid**. The people who use the electrical energy are called consumers (Figure 10.7).

The National Grid links together all the major power stations as well as smaller electrical energy generators. This means that power stations can be closed for maintenance without disrupting the supply of electricity. Some power stations are used to supply energy to the grid only at times of peak demand. (Imagine the extra demand at half time in a televised world cup football match. Everyone wants a cup of tea or coffee.)

Figure 10.8 Pylons carry the main cables of the National Grid to all parts of the country.

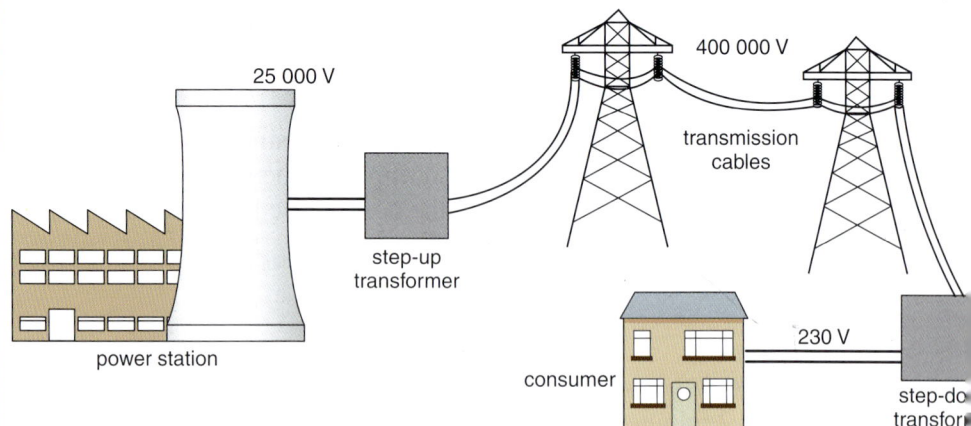

Figure 10.7 A simplified diagram of the National Grid network

Every second, huge amounts of electrical energy are transmitted through the National Grid from power stations to consumers. The energy is transmitted using a high voltage. Using a **transformer** to increase the voltage across the cables reduces the current through the cables. Reducing the current reduces the amount of electrical energy transformed into heat in the cables. By reducing the energy lost as heat in the cables, the transmission of mains electricity is made more efficient. So, more of the energy from the power station gets to the consumer.

10.4 How should we generate the electricity we need?

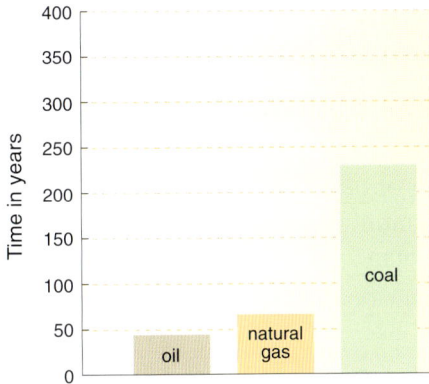

Figure 10.9 How long will the reserves of fossil fuels last?

The most common energy sources for generating electricity are coal, oil and natural gas. Coal, oil and natural gas are **fossil fuels**. Like all fuels, they store energy. But to release their energy, the fuels must be burned. It would take millions of years to replace the fossil fuels that have already been used. Because of this, fossil fuels are described as **non-renewable energy sources**. Once they are gone, they are gone forever.

Fossil fuel reserves are limited. Although some companies are always trying to find new reserves, fossil fuels will eventually run out. Figure 10.9 shows how long we can expect fossil fuels to last if we continue to use them at the present rate.

If we want the Earth's reserves of fossil fuels to last longer then we must start to use them more efficiently. This is crucial for oil and natural gas.

In comparison to fossil fuels, resources like the wind and the tides will never run out. These are energy sources that are replaced as fast as they are used. Because of this, the wind and the tides are described as **renewable energy sources**.

Although both non-renewable and renewable energy sources are used to generate electricity, most electricity in Britain is generated in power stations from non-renewable fossil fuels.

Coal or oil is burned to heat water and produce steam. The steam is made to drive turbines. The turbines turn generators and the generators produce electricity (Figure 10.10).

In a gas-burning power station there is no need to produce steam. Heat from the burning gas produces fast moving hot air which is used to drive the turbines directly.

8 Before the National Grid was set up in 1926, each area in Britain had its own power station. What are the advantages of the National Grid rather than smaller areas generating their own electricity?

9 Transformers are usually very efficient. Explain what this means.

10 Why does the National Grid transmit electrical energy at high voltage?

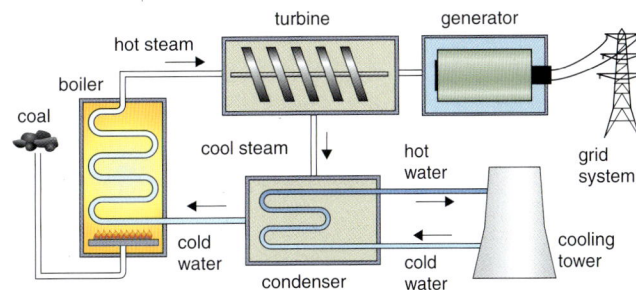

Figure 10.10 Using coal or oil to generate electricity

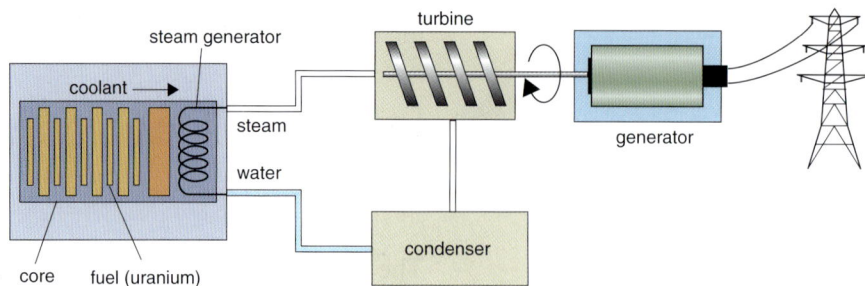

Figure 10.11 Using nuclear fuel to generate electricity

The breaking up or splitting of large atoms in elements like uranium and plutonium is called **nuclear fission**. When nuclear fission happens, energy is released.

Nuclear power stations generate electricity in a similar way to those that burn coal. But the fuel, mainly uranium or plutonium, is not burned. Instead the heat given out when the uranium or plutonium atoms break up is removed from the reactor by a coolant, and then used to turn water into steam (Figure 10.11).

10.5 Why are non-renewable fuels used to generate electricity?

Power stations that use non-renewable fuels can generate electricity at any time. It doesn't matter whether it is day or night, summer or winter. Provided the fuel keeps arriving the power station keeps on generating. This makes non-renewable fuels reliable energy sources.

Oil- and coal-burning power stations are relatively cheap to build. Most of them have a very high power output of about 2000 megawatts (that's 2000 million watts). So a small number of power stations can provide power to millions of consumers.

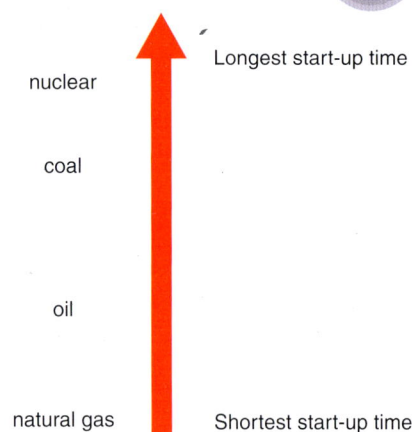

Figure 10.12 Some types of power station start up faster than others

After a power station has been closed down for maintenance, it needs to be started up again. The time it takes to get started up and generating electricity depends on the type of power station (Figure 10.12).

Coal-burning and most oil-burning power stations need to be kept running all the time. This is because the furnaces are likely to be damaged if they are allowed to cool down.

11 How many 4 MW wind turbines are needed to replace a 2000 MW coal-burning power station?

12 Give one reason why gas-burning power stations start up quicker than coal-burning power stations.

Nuclear fuel, which is relatively cheap, provides a concentrated energy source. One kilogram of nuclear fuel can release the same amount of energy as 20 000 kilograms of coal. The downside is the very high cost of building and safely decommissioning (dismantling) nuclear power stations.

How does the use of non-renewable fuel affect the environment?

Fossil fuels and global warming

Most of the heat (infra-red radiation) from the Sun that hits the Earth's surface is absorbed. Some of it is radiated back into space.

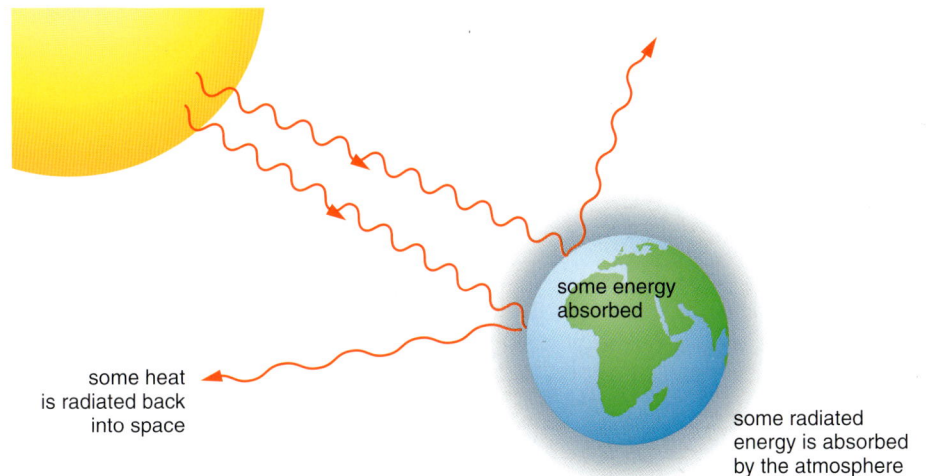

some energy absorbed

some heat is radiated back into space

some radiated energy is absorbed by the atmosphere

Figure 10.13 Some of the Sun's heat is radiated back into space

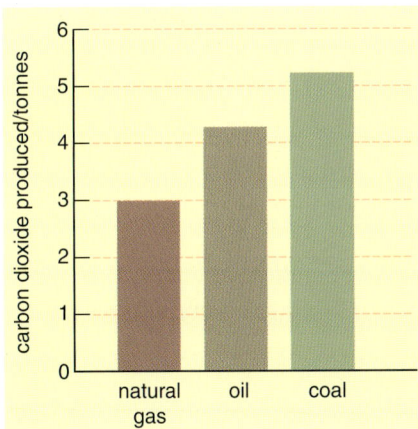

The most penetrating heat rays from the Sun pass through the atmosphere and warm the Earth. These rays are absorbed by the Earth, which, in turn, radiates less penetrating heat rays. Under ideal conditions, the heat received from the Sun equals the heat radiated away from the Earth, and the average temperature of the Earth stays constant. But, during the last 150 years, the concentration of carbon dioxide in the Earth's atmosphere has increased. Heat rays cannot penetrate carbon dioxide as well as other gases in the air and, because of this, the less penetrating radiation from the Earth is blocked by carbon dioxide and prevented from travelling into space.

Figure 10.14 The amount of carbon dioxide produced for each kilowatt of power generated using a fossil fuel

Fuel-burning power stations produce a lot of waste gas. The main constituent of this waste gas is carbon dioxide. So, as more and more fossil fuels are burned, more carbon dioxide enters the atmosphere. Many people believe that this has led to more heat being trapped and an increase in worldwide temperatures. This increase in worldwide temperatures is called **global warming**. There is clear evidence that the polar ice caps are melting due to this increase in temperature.

Fossil fuels and acid rain

When coal and oil are burned, sulfur dioxide is produced. This sulfur dioxide ultimately results in acid rain which damages buildings, kills plants and pollutes rivers and lakes (see Section 6.4).

Figure 10.15 Trees killed by acid rain

Figure 10.16 Oil doesn't have to be burned to affect the environment. An oil spill from a tanker can have a serious effect on wildlife and the environment.

⑬ Explain why the use of biofuels to generate electricity would benefit farming communities and the environment.

Acid rain can be reduced.
- Sulfur can be removed from the fuels before they are burned.
- Sulfur dioxide can be removed from the waste gases before they enter the atmosphere.

Both methods of reducing acid rain are expensive and add to the cost of the electricity generated. Equipment to remove sulfur dioxide from the emissions at Drax, the UK's largest power station, cost 680 million pounds!

Into the future

Like fossil fuels, biofuels such as woodchips, palm nuts, straw and olives also emit carbon dioxide when they burn. But unlike fossil fuels, they use up carbon dioxide as they grow. So, overall they add no extra carbon dioxide to the environment. Some power stations are now experimenting with blends of biofuel and coal. Elean Power Station in Cambridgeshire burns straw to generate 36 megawatts of power each year. In Queensland, Australia, a local electricity supplier burns 5000 tonnes of nut shells every year, producing enough electricity for 1200 homes.

Nuclear fuels and the environment

Nuclear fuels do not produce carbon dioxide or sulfur dioxide. Consequently, they do not add to global warming or produce acid rain.

Some people worry that nuclear power stations will leak radiation, but when nuclear power stations are working normally, little or no radiation or radioactive materials should enter the environment.

Fortunately, serious accidents at nuclear power stations are rare. But, when they do occur, radiation can be carried by the wind to a very large area.

Nuclear power stations do produce radioactive waste. This must be stored safely for long periods of time, sometimes for thousands of years.

Career – health physicist

Health physicists are responsible for monitoring radiation in power stations and the surrounding area. It's their job to make sure that people working at the power station are exposed to as little radiation as possible. They also monitor the local environment to ensure that the power station is not causing levels of radiation to rise above normal.

Activity – Nuclear waste? Not in our backyard!

Suppose you are the scientist on a committee asked to recommend the best way to deal with nuclear waste. The committee has put forward three possible options. All involve burying the waste.

Option 1:   Bury and seal the waste permanently 500 m underground.

Option 2:   Bury the waste 500 m underground but continue to monitor and, if necessary, retrieve the waste.

Option 3:   Bury waste with a short half-life 100 m underground.

Some people don't want the area they live in to be chosen as a site for burying nuclear waste. They have started a campaign to prevent the waste being buried on their doorstep.

❶ What arguments might the campaigners use against burying nuclear waste in their area?

❷ What arguments and evidence would you use to persuade people that burying nuclear waste is a safe option?

❸ Explain why the committee has suggested that waste with a short half-life could be buried only 100 m below ground.

Look at Section 11.11 if you need help with this activity.

---

⑭ How is the production of electricity in a coal-fired power station different to that in a nuclear power station?

⑮ What economic factors need to be considered before building a nuclear power station?

⑯ A 2400 MW power station uses nuclear fuel and 1 kg of the nuclear fuel produces 1 600 000 kWh of electrical energy.
   a) Calculate the power of the station in kilowatts (kW).

   b) How many kilowatt-hours (kWh) of electrical energy does the power station produce every hour?
   c) Calculate the amount of nuclear fuel the power station uses in one hour.

⑰ Richard is considering buying some land to build a house on. The land is close to a nuclear power station. What evidence could a health physicist working at the power station present to Richard, to persuade him that it would be safe to live in a house built on the land?

10.6 Why use renewable energy sources to generate electricity?

Non-renewable fuels, like oil and gas, will not last forever. Reserves are running out. But the demand for electricity increases at an ever-increasing rate. Renewable energy sources need not burn fuels to generate electricity. The turbines that turn the generators use energy direct from the renewable source.

At the moment, Britain produces less than 3% of its electricity from renewable energy sources. The target, set by the government, is to increase this to 25% by the year 2025. But, how will this target be met?

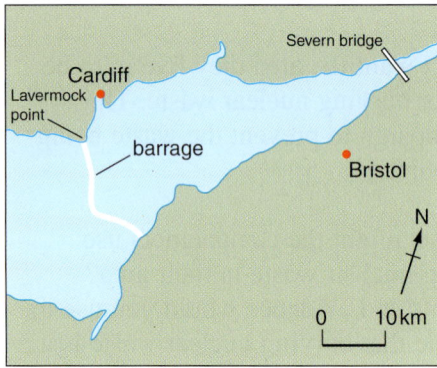

Figure 10.17 A tidal barrage across the River Severn could supply about 6% of Britain's electricity.

Using the tides

Every day tides rise and fall. Massive amounts of water move in and out of river estuaries. It is estimated that the energy of the tides could generate up to 20% of Britain's electricity.

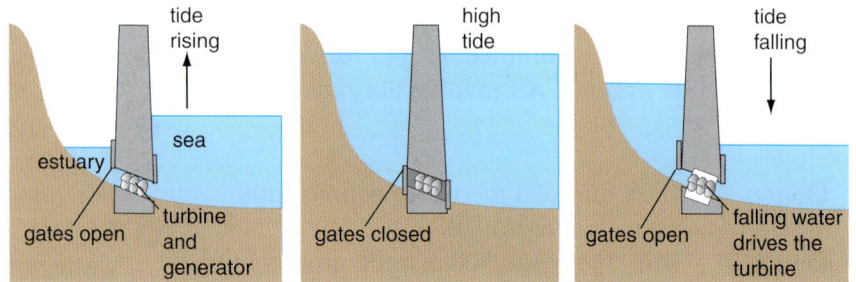

Figure 10.18 How a tidal barrage works

A barrage, which is like a dam, is built across a river estuary (Figure 10.18). The barrage has underwater gates which open as the tide comes in and then close to trap the water behind the barrage. When the tide goes out, a second set of gates are opened. Water rushes through these gates, driving turbines. The turbines turn generators which produce electricity. In some cases this system can be made even more efficient by also using the incoming tides to drive the turbines.

Of course, the amount of electricity produced by this method depends on the tides. The tides vary each day and change on a monthly cycle. However, tides are predictable, so the output from a tidal generator is reliable.

Tidal barrages have to be built across river estuaries. Because of this, they can disturb the flow of the river. This may destroy the habitats of wading birds and the mud-living organisms on which they feed.

Using waves

The UK is surrounded by the sea and the potential for using wave energy to generate electricity is enormous. So, why are there so few wave power stations? The answer is simple. It is very difficult to harness wave energy and transform it into large amounts of electricity.

LIMPET (Figure 10.19), built on the coast of Islay, is an oscillating water column generator. The movement of the Atlantic waves forces air to drive a turbine, which then turns a generator (Figure 10.20).

Figure 10.19 LIMPET – the world's first commercial wave power station on the Scottish island of Islay has been generating electricity since November 2000.

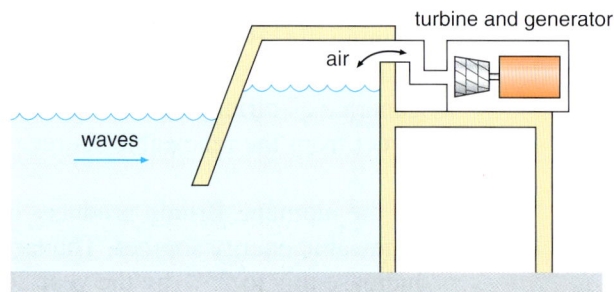

Figure 10.20 An oscillating water column generator

Activity – Generating electricity: the future

Because of the expense involved, scientists design and build smaller versions of their products for initial testing. These are called prototypes. After trials, they modify these prototypes before going ahead with the construction of a final product.

❶ Use the internet to find out how the two generator prototypes, Seaflow and Pelamis, work. Start your search by looking at the following websites: www.marineturbines.com   www.bwea.com   www.oceanpd.com

❷ Write a few sentences about each generator prototype.

❸ Which of the two prototypes do you think is the best choice for the future? Explain your choice.

Figure 10.21 The Seaflow generator, seen here, has been raised out of the water for maintenance

Figure 10.22 Pelamis, moored off the Orkney Islands, moves with the motion of the waves

Using hydroelectric power

The energy of a river can be used to generate electricity. To do this, a dam must be built across the river. Water is then trapped forming a lake behind the dam. When the trapped water is released it rushes downhill. The gravitational potential energy of the falling water is used to drive turbines, which then turn generators.

Hydroelectric power stations generate about 2% of Britain's electricity, but they produce about 10% of the world's electricity. Many of these schemes involve flooding large areas of land behind a dam. This may mean that:
- forests are cut down;
- farmland is lost;
- wildlife habitats are destroyed.

In China, the world's largest power project involved building a dam across the Yangtze River. The lake behind the dam flooded so much land that 1.5 million people had to move to new homes.

Figure 10.23 Hydroelectric power stations can be very large in size and generate vast amounts of power

The demand for electricity changes during the day. When demand is high a hydroelectric pumped storage system can provide the extra electricity (Figure 10.26).

In just a few seconds, water which has been pumped into the top lake can be released. As the water falls to the bottom lake it drives a turbine, which then turns a generator.

At night, when more electricity is being generated than is needed, electricity is used to pump water back into the top lake. So, the energy has been stored and the power station is ready to generate again when demand is high.

Figure 10.24 Between 1964 and 1968 the ancient temple of Abu Simbel in Egypt was moved, stone by stone, to higher ground. This prevented it being flooded by the lake created behind the Aswan High Dam.

18 What environmental problems are caused by building hydroelectric power stations?

19 Why do pumped storage power stations not increase energy resources?

Figure 10.25 Ffestiniog Power Station is the first major pumped storage power station in the UK. It can produce enough electricity to keep North Wales going for several hours.

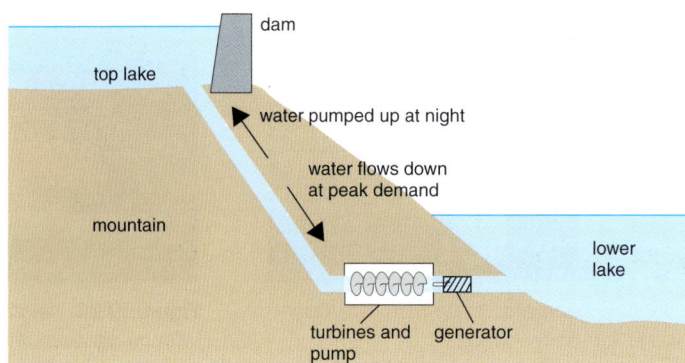

Figure 10.26 A pumped storage power station

Activity – Investigating a water-powered electricity generator

The model generator investigated by Anya is shown in Figure 10.27. The turbine rotates when a stream of water hits the cups. The rotating turbine turns the generator, which produces electricity.

Anya wanted to find out if the voltage output from the generator depended on the volume of water hitting the turbine each second.

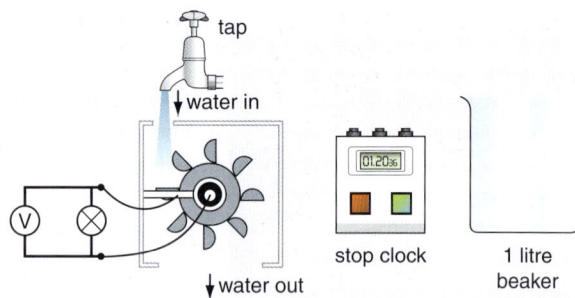

Figure 10.27 The apparatus used by Anya to investigate the generator

Having made sure that the turbine was directly under the tap, Anya slowly increased the flow of water until the turbine just started to turn. Anya then placed a beaker in the water flow, timed one second, and then removed the beaker. Anya measured the volume of water and then tipped it away before repeating this step twice more. Her results are recorded in Table 10.2.

Attempt number	Volume of water collected (ml)
1	42
2	52
3	68

Table 10.2 Anya's first set of results

1 What is the mean (average) volume of water collected?

2 Suggest a reason for the big difference between these values.

Juspal suggested that the experiment would be more **accurate** if Anya collected the water for 10 seconds rather than one second. She could then divide this by 10 to get the volume of water flowing in one second.

3 Why would the method suggested by Juspal improve the accuracy of the investigation?

Anya took the following steps to complete her investigation.
1 Turn the tap to give a steady flow of water.
2 Record the voltmeter reading.
3 Collect the water from the tap for 10 seconds.
4 Increase the flow of water and repeat steps 2 and 3.

The results of Anya's investigation are presented in Table 10.3 and Figure 10.28.

Figure 10.28 A graph of Anya's results

Voltage (V)	Volume of water collected in 10 s (ml)	Volume of water collected in 1 s (ml)
0.25	680	68
0.30	720	72
0.50	780	78
0.60	880	88
0.70	940	94

Table 10.3 The results of Anya's investigation

4 Anya only took one set of results. Suggest why it would have been difficult to repeat the experiment to obtain a comparable set of results.

5 Do you think Anya planned a fair test? Give reasons for your answer.

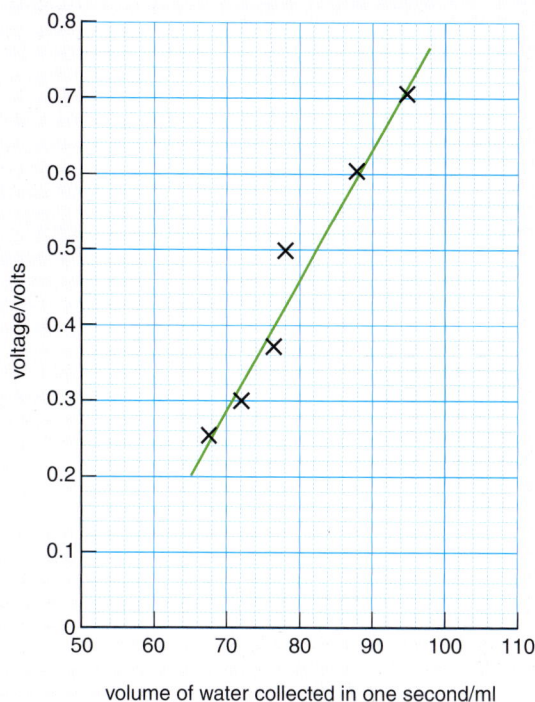

Not all of Anya's results lie on a straight line. What has been drawn is the **line of best fit**. Notice that some of the points are on one side of this line and some on the other. If all the points are to one side of the line it is *not* the line of best fit.

6 Suggest why the results do not give a graph with a perfect straight line.

The results of the investigation show a relationship or pattern between the volume of water hitting the turbine each second and the voltage generated. However the graph clearly shows that these two quantities are not **directly proportional**. For two quantities to be directly proportional the graph must give a straight line passing through the origin, the point (0,0).

7 Describe the pattern linking the volume of water hitting the turbine each second and the voltage generated.

8 Estimate the volume of water needed to generate 0.8 V. To do this you must assume the pattern continues as a straight line. This is called **extrapolating** the results.

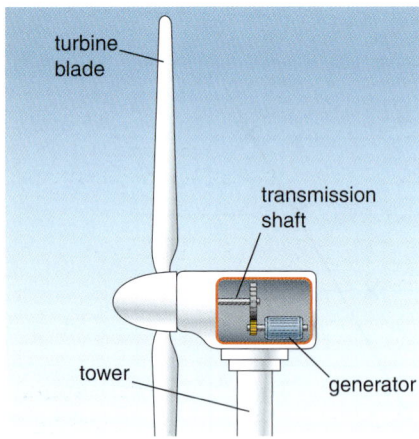

Figure 10.29 Inside a wind-powered generator

Using the wind

Using energy from the wind is not a new idea. Over 1000 years ago, windmills were used to grind grain. In some countries, the power of the wind is still used to pump water from natural underground reservoirs.

In a wind generator, the turbine blades rotate when the wind blows. This turns the generator and produces electricity. So, a wind turbine transforms the kinetic energy of the wind to electrical energy (Figure 10.29).

One problem with wind generators is that the amount of electricity generated changes with the strength of the wind. If the wind is too light, little or no electricity is produced. This makes the wind an unreliable energy resource.

Wind energy applications vary from small units, which charge a battery, to large commercial **wind farms** with large numbers of wind generators providing power to the National Grid. Currently, only about 1% of Britain's electricity is generated using the wind, but this could increase to about 10%.

places where there is usually a strong wind

20 Why are the tides a more reliable way of generating electricity than the wind?

21 Look at Figure 10.32. Why do you think the cost of generating electricity using wind turbines has fallen?

22 The inhabitants of a small island need to replace their old power station. Half the residents want to build a coal-burning power station. The other half want a wind farm.
a) State the reasons for and against a coal-burning power station.
b) State the reasons for and against a wind farm.

Figure 10.30 Using a wind generator to charge the batteries of a boat

Figure 10.31 Generating electricity from the wind can be made more reliable by putting the wind turbines in places where there is usually a strong wind

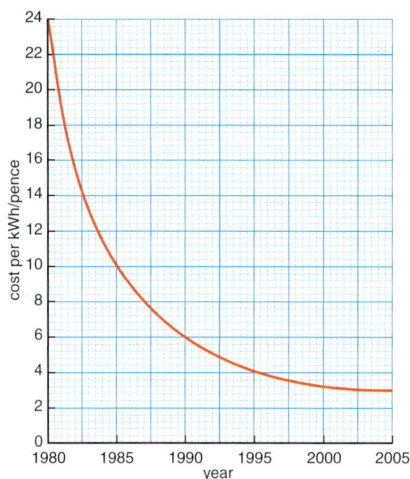

Figure 10.32 The cost of generating electricity using wind turbines

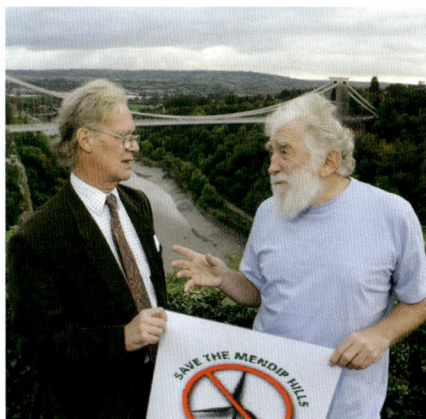

Figure 10.33 The conservationist, Professor David Bellamy voices his opposition against placing the first giant wind turbine on the Mendip Hills in Somerset

Activity – Using wind power at home

You have decided to generate your own electricity using wind power. Before starting you need to know if this is a practical idea. Use the internet to research the information you need before deciding to go ahead. Use the British Wind Energy Association website (www.bwea.com) to help you.

Many people, including scientists, engineers and conservationists, argue that wind turbines are costly and inefficient. They also have a negative effect on the environment and wildlife.

Wind farms need to be where it is windy! This is usually on hills or along the coast in places of natural beauty. Some people think that the wind farms are ugly and spoil the view. So to some people, wind farms cause visual pollution.

Figure 10.34 Windy places are often in places of natural beauty

People living near wind generators also find them noisy. There is low-frequency noise from the rotating turbine blades and noise from the machinery inside the generator. So for some people wind farms cause noise pollution and, in certain cases, stress-related illness.

The costs involved in constructing a wind farm should not be ignored. Apart from the cost of making each turbine, materials are needed to build the base for each turbine and the access roads to and from them. Each base, for the largest type of turbine currently being manufactured, requires about 1000 tonnes of concrete. The production of cement for this amount of concrete will, in itself, pollute the environment. (See Section 5.5.)

Plans to site wind turbines offshore would greatly increase the amount of electricity generated by wind power in the UK. So far, three areas in the UK have been approved as sites for offshore wind farms. One of these, at Scroby Sands off the Norfolk coast consisting of 39 giant turbines, will generate electricity for 50 000 homes.

Figure 10.35 An offshore wind farm

Activity – For or against wind farms

The strength of local opposition to a planned wind farm of 26 giant turbines on a stretch of coastal marshland, near the Essex villages of Bradwell-on-sea and Tillingham, has led to the project being stopped.

People living in these villages and scattered farms were appalled by the plans. They said the plans would:

- destroy the unspoilt coastal marshes;
- threaten the thousands of migrating wildfowl and wading birds;
- create high levels of noise pollution from the turbines.

❶ Imagine you are a scientist working for a wind power company. What arguments would you use to persuade people to accept a wind farm in their area?

❷ Do scientists working for commercial organisations and environmental groups only look for evidence to support their own views? Give reasons for your answer.

10.7 Energy from the Sun and Earth

Energy direct from the Sun

The amount of electricity produced by a solar cell depends on the energy of the sunlight falling on it. If it's dark or cloudy, the energy of the light falling on the cell will be low and little or no electricity will be produced. Solar cells are therefore an unreliable way of producing electricity.

In some situations solar cells are the most convenient way to produce electricity. They are often used in remote areas, on satellites and in devices where only a small amount of electricity is needed.

Figure 10.36 A solar-powered MP3 player

Figure 10.37 Solar cells produce the electricity to operate this satellite

㉓ Why are solar cells suitable for use with
  a) a mobile phone charger?
  b) a satellite?

㉔ Each square metre of land in the UK receives, on average, 200 joules of energy per second from the Sun.
  a) What area of land would need to be covered by solar cells in order to generate 1 MW of power? Assume that solar cells transform only 15% of solar energy into electrical energy.
  b) Would it be sensible to build a solar power station in the UK? Explain your answer.

㉕ An African village uses solar cells to generate the electricity needed to operate the pump at a water well. Why are solar cells used to generate the electricity, rather than a petrol generator?

Energy from the Earth

Uranium and other radioactive elements are found inside the Earth. When atoms of these radioactive elements decay (Sections 11.8 and 11.9), they transfer heat to the surrounding rocks. This is called **geothermal energy**.

In areas of volcanoes, geysers and hot springs, water and steam heated by geothermal energy often reaches the Earth's surface. The steam can be used to drive turbines connected to electricity generators.

Although the world's first geothermal power station, in Italy, started generating electricity in 1904, geothermal energy is under used. At present the total worldwide geothermal generating capacity is only 8400 MW. But, this is gradually changing as more countries invest in geothermal energy. In the Phillipines, one quarter of all electricity is now generated using geothermal energy.

Figure 10.38 Letting off steam the natural way!

26 The graph in Figure 10.39 shows how the demand for electricity in the UK varies over 24 hours.
   a) Why does demand increase after 6.00a.m?
   b) The average demand during a 24-hour period is 6000 MW. How many hours was the actual demand lower than this?
   c) What type of power station could supply the additional power needed at peak times?

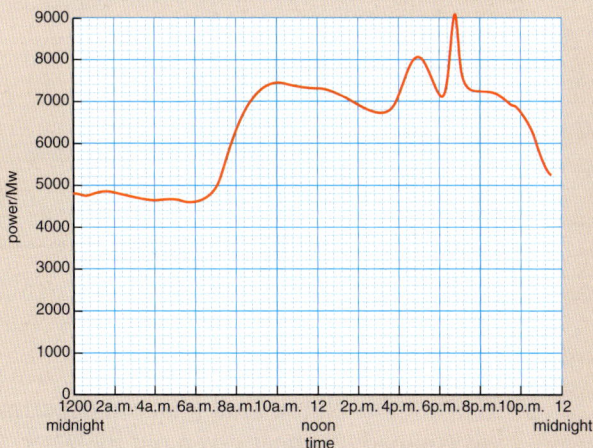

Figure 10.39 The demand for electricity is not constant

27 Each year the global demand for electricity grows (Figure 10.40). How do you think the increased demand for electricity can be satisfied?

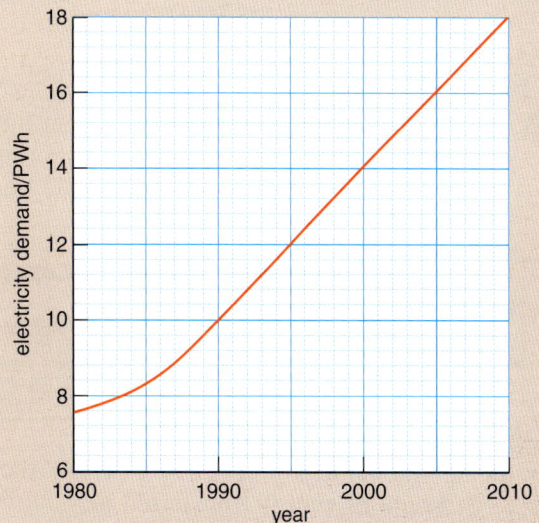

Figure 10.40 The actual demand for electricity and the forecasted demand for electricity between 1980 and 2010

✓ **Electrical energy** is easily transformed into other forms of energy.

✓ **Power** is the rate at which a device transforms energy.

✓ The power of a device is measured in watts (W) or kilowatts (kW).

✓ 1 watt (W) = 1 joule / second (J/s).

✓ The amount of electrical energy that a device transforms depends on the power of the device and how long it is switched on for.

✓ The amount of electricity transferred from the mains can be calculated using the equation:

energy transferred   =   power   ×   time
(in kilowatt-hours)   (in kilowatts)   (in hours)

✓ The **National Grid** is a network of cables and transformers linking power stations to consumers.

✓ Non-renewable and renewable energy sources can be used to generate electricity.

✓ The most common energy sources are fossil fuels – coal, oil and natural gas. Fossil fuels are non-renewable energy sources.

✓ Renewable energy sources include the wind, tides, waves, falling water in hydroelectric schemes, solar energy from the Sun and geothermal energy from hot rocks.

✓ In most power stations an energy source is used to heat water. The steam produced drives a turbine, which turns an electrical generator.

✓ The energy from renewable sources can be used to drive turbines directly.

✓ Each type of energy source has its advantages and disadvantages.

EXAM QUESTIONS

❶ Table 10.4 shows data about a filament light bulb and a low-energy light bulb. Although the bulbs have different power ratings they both emit the same amount of light.

Type of bulb	Power (w)	Lifetime (hours)	Cost (£)
Filament	100	1 000	0.60
Low energy	20	12 000	1.85

Table 10.4

a) Calculate, in kilowatt-hours (kWh), the amount of energy transferred to:
   i)   the filament bulb during its lifetime;
   ii) the low-energy bulb during its lifetime.
   (2 marks)
b) If each kilowatt-hour (kWh) of energy costs 12p, calculate the cost of using each bulb for its lifetime. *(2 marks)*
c) Which of the two bulbs is the most cost-effective? To gain full marks you must justify your answer with a calculation.
   (2 marks)

❷ An advertisement for solid fuel firelighters claims:

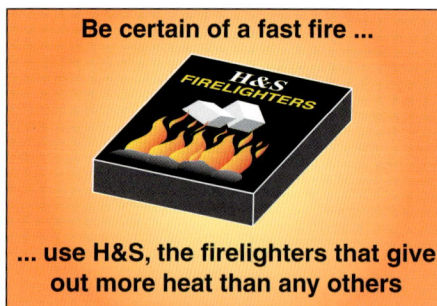

Figure 10.41 An advertisement produced by a manufacturer

a) To test this claim, Paul plans an experiment to compare the heat given out by H&S firelighters with two other brands.
   This is Paul's plan for the experiment.
   • Take 1 g of H&S firelighter and place it on a tin lid.
   • Put 80 ml of water into a beaker.
   • Measure the temperature of the water.

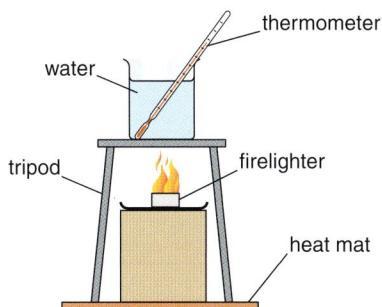

Figure 10.42 The apparatus used in Paul's experiment

- Use a match to set fire to the firelighter and then use the burning firelighter to heat the water.
- When all the firelighter has burned, measure the new water temperature.
- Repeat the experiment with the other two brands of firelighter.
    i)   The type of firelighter is a variable. Is it a categoric or continuous variable? *(1 mark)*
    ii) Name two variables that Paul kept the same. *(2 marks)*
    iii) Give one reason why Paul wore safety goggles during the experiment. *(1 mark)*
    iv) Suggest how Paul could have improved the way in which he collected the data. *(2 marks)*
    v)   Suggest one change Paul could have made to his choice of measuring instruments that would have improved the accuracy or precision of the experiment. *(1 mark)*

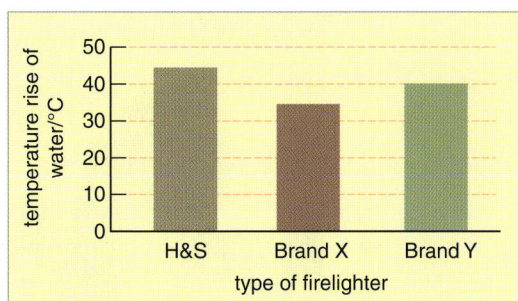

Figure 10.433 The data from Paul's experiment is displayed in this bar chart

b) To compare his data, Paul drew the bar chart shown in Figure 10.43.
    i)   Was the data collected by Paul sufficient to confirm the claim made by the maker of H&S firelighters? Give a reason for your answer. *(2 marks)*
    ii) Give two reasons why not all of the heat produced by the burning firelighter was transferred to the water. *(2 marks)*

3 Mary wants to generate the electricity needed to run her own home. Having considered various options, including a petrol generator, she has decided to have solar panels installed on the roof of her house.

a) Use the following data to calculate the area of solar panels Mary needs to meet her electrical needs. Give your answer to the nearest whole number. *(3 marks)*

Electrical energy needed each year	3200 kWh
Average solar energy collected each day by 1 m² of solar panel	4 kWh
Efficiency of solar panel	20%

b) Explain why Mary should not rely only on solar panels to generate the electricity she needs in the winter. *(2 marks)*

c) Why should Mary keep her house connected to the National Grid? *(2 marks)*

d) Give one advantage of using a petrol generator rather than solar panels to generate electricity. *(1 mark)*

Chapter 11
What are the properties, uses and hazards of electromagnetic waves and radioactive substances?

At the end of this chapter you should:

✓ know that electromagnetic radiations travel as waves and move energy from one place to another;

✓ know that electromagnetic waves are grouped (classified) according to wavelength and frequency;

✓ understand that electromagnetic radiations and radioactive substances have many useful applications;

✓ appreciate that there are hazards associated with the uses of electromagnetic and nuclear radiations;

✓ understand how to reduce exposure to different types of electromagnetic and nuclear radiation;

✓ know that communication signals can be analogue or digital;

✓ know that all atoms have a small central nucleus, composed of protons and neutrons, surrounded by electrons;

✓ know that radioactive substances emit three main types of radiation (alpha particles, beta particles and gamma rays) from the nuclei of their atoms;

✓ understand the nature and important properties of alpha particles, beta particles and gamma rays;

✓ understand the term 'half-life'.

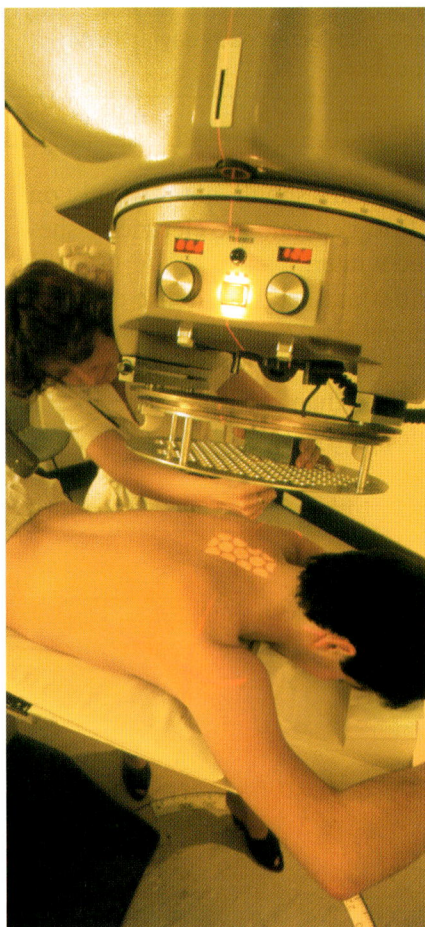

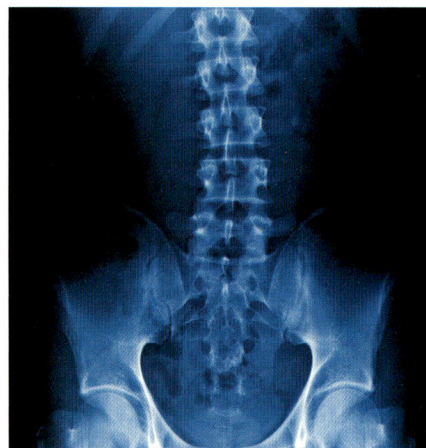

Figure 11.1 Some of the uses of electromagnetic waves and radioactive substances involve potential hazards as well as obvious benefits

11.1 Looking at waves

A good way to visualise a wave is to use a stretched 'slinky' spring. In Figure 11.2 two people have stretched a slinky across the floor. While one end of the slinky is held still, the other end is moved from side to side. This sends a series of wave pulses along the slinky. The person holding the still end of the slinky can feel the pulses as they arrive. Each pulse carries energy but when the pulse has passed, the slinky remains exactly as it was before. None of the material of the slinky has moved permanently.

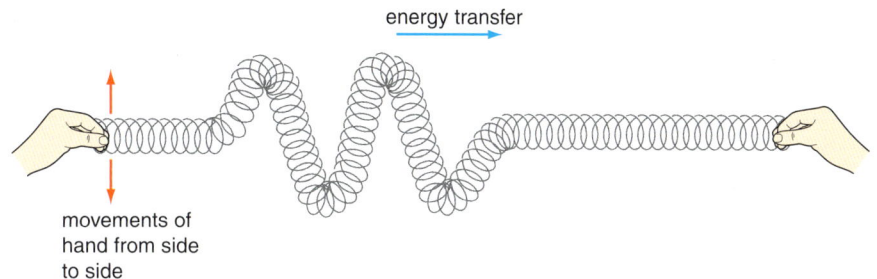

energy transfer

movements of hand from side to side

Figure 11.2 Wave pulses moving along a slinky

So, waves move energy from one place to another without transferring any material (matter).

This is just like a Mexican wave at a sports event. As the Mexican wave rushes around the stadium you get the impression of something moving, but the spectators stay where they are.

When raindrops fall into a pond, water ripples spread out from the point where the rain hits the water. Again, you get the impression of something moving along but it is *not* the water. Energy is being transferred while the water vibrates up and down. You can see this clearly if you watch a cork bobbing up and down on the surface of a pond as waves pass it by (Figure 11.4).

All waves can be described in terms of three quantities – **wavelength, frequency** and **amplitude**. Being able to see a wave, like a water wave, makes it much easier to understand what these quantities are (Figure 11.5).

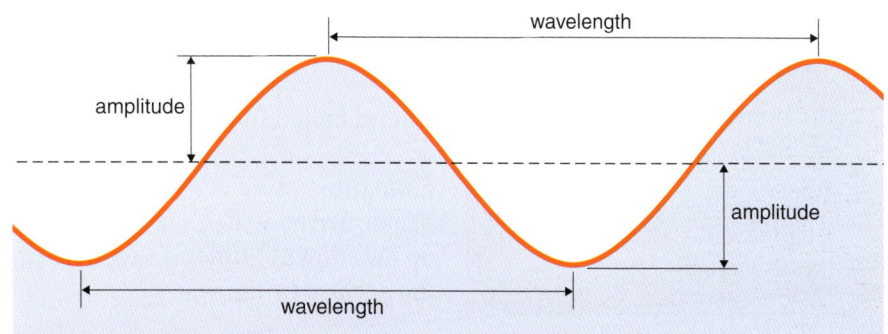

Figure 11.3 A Mexican wave passes around a stadium

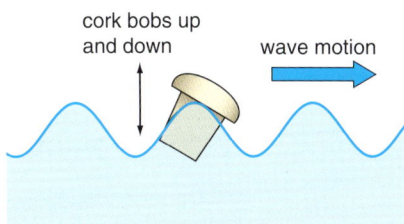

cork bobs up and down

wave motion

Figure 11.4 The cork bobs up and down as the water waves pass it by

Wavelength is the distance from a point on one wave to the equivalent point on the next wave.

Frequency is the number of waves produced each second. It is also the number of waves that pass a point each second.

wavelength

amplitude

amplitude

wavelength

Figure 11.5 A wave moving across a water surface

Frequency is measured in units called **hertz (Hz)**. A source producing one wave every second has a frequency of one hertz (1 Hz). If there are 50 waves in 10 seconds, the frequency is five waves per second, or 5 Hz.

The amplitude depends on the energy of the wave. The greater the amplitude, the greater the energy that the wave is transferring.

Wave speed and the wave equation

Suppose you are watching waves on the surface of a pond and three waves pass a particular point in one second (frequency, f = 3 Hz). Suppose also that the wavelength is 2 cm (Figure 11.7).

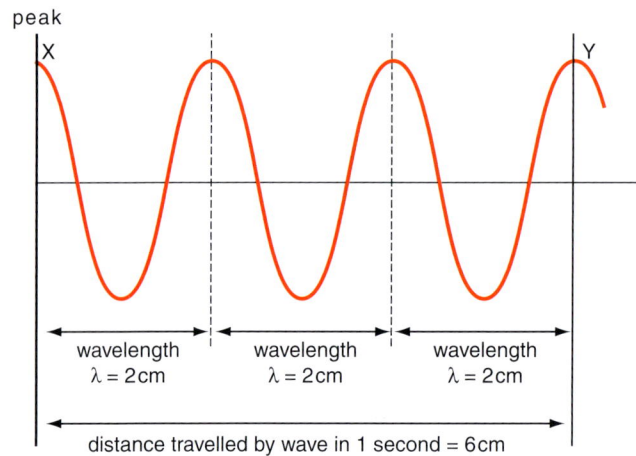

Figure 11.7

Each wave peak moves forward by three complete waves every second. So, in one second the peak at X will have moved to Y. So the distance moved by a peak in one second is 6 cm.

Therefore, the speed of the wave = 6 cm/s.

In Figure 11.7 the speed of the waves can also be calculated by multiplying the number of waves per second (the frequency) by the length of each wave (the wavelength), i.e.

$$\text{wave speed} \atop \text{(in metres/second)} \quad = \quad \text{frequency} \atop \text{(in hertz)} \quad \times \quad \text{wavelength} \atop \text{(in metres)}$$

This is called the **wave equation** and it applies to all waves.

Example
Water waves with a wavelength of 0.6 metres make a moored boat bob up and down 2 times a second. At what speed do the waves travel across the surface of the water?

$$\text{wave speed} = \text{frequency} \times \text{wavelength}$$
$$\text{wave speed} = 2 \times 0.6 = 1.2 \text{ m/s}$$

Amplitude is the maximum displacement of a wave from its mean (middle) position.

❶ Figure 11.6 shows wave pulses travelling at the same speed along two elastic cords. How are the wave pulses travelling along A different to those travelling along B?

cord A

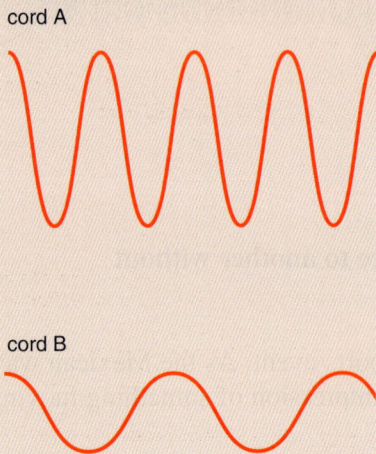

cord B

Figure 11.6

❷ The wave maker at a leisure pool makes 1 wave every 2 seconds. The wavelength of each wave is 4 metres.
   a) What is the frequency of the waves?
   b) Calculate the speed that the waves travel across the surface of the water.

❸ When a stone is thrown into a pond, waves spread out across the pond. The speed of the waves is 0.3 m/s and their wavelength 5 cm.

Calculate the frequency of these waves.

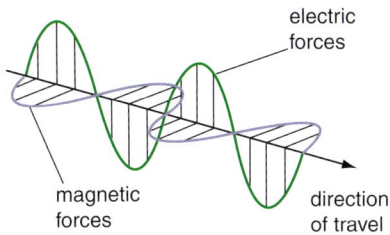

Figure 11.8 Vibrating electric and magnetic forces

The electromagnetic 'family'

Why electromagnetic?

Electromagnetic radiation travels as waves. But it is hard to visualise an electromagnetic wave. Just like water waves they transfer energy. But unlike water waves they do it without the need for a medium (a material to move through). The energy is carried through space by vibrating electric and magnetic forces moving along together. This is what we call an **electromagnetic wave**.

The electromagnetic spectrum

You already know a lot about one part of the **electromagnetic spectrum** – the part we call visible light. However, visible light is only a small part of the much larger electromagnetic 'family'. The electromagnetic family is a whole 'family' of waves that together make up the electromagnetic spectrum.

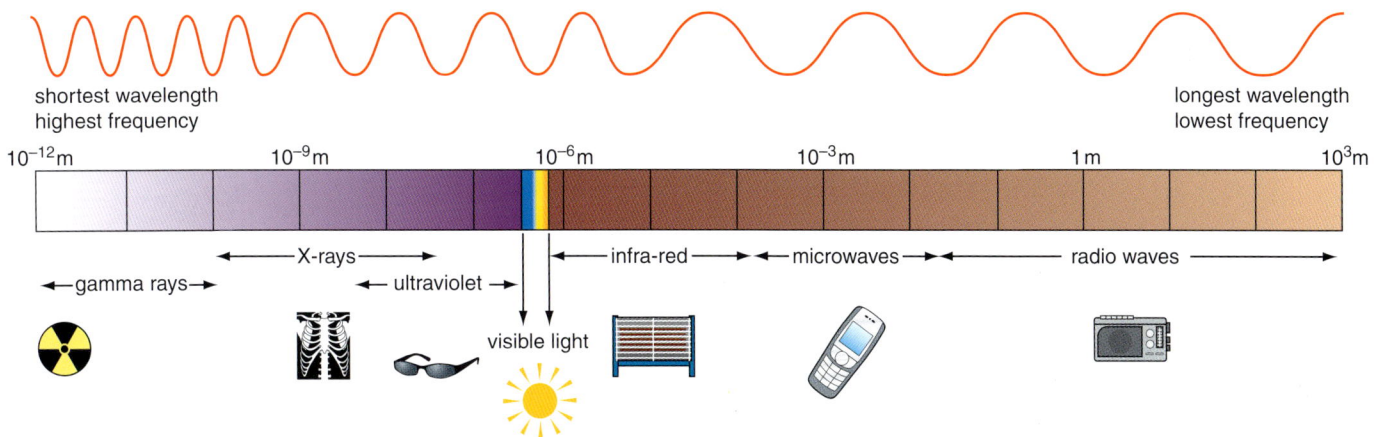

Figure 11.9 The electromagnetic spectrum – a 'family' of waves

4 Which parts of the electromagnetic spectrum have a higher frequency than ultraviolet?

5 A navigation system transmits radio waves at a frequency of 1.5 MHz.
   a) What is the speed of the radio waves?
   b) Calculate the wavelength of the radio waves.
      (1 MHz = 1 000 000 Hz.)

There are no gaps in wavelength in the electromagnetic spectrum. Wavelengths run smoothly from one value to another, forming a continuous spectrum.

Despite having a huge range of wavelengths and frequencies, all electromagnetic waves have some important properties in common.
- They obey the wave equation (wave speed = wavelength × frequency).
- They can travel through a vacuum.
- They all travel through a vacuum and air at the same speed of 300 000 000 m/s. (This means that waves with the lowest frequency have the longest wavelength and waves with the highest frequency have the shortest wavelength.)
- They can all be reflected, absorbed or transmitted. But different wavelengths of electromagnetic radiation are reflected, absorbed or transmitted to different extents by different materials or types of surface.

6 The waves which carry Radio 4 on the long-wave band have a wavelength of 1500 metres. Calculate the frequency of these carrier waves.

Electromagnetic radiation carries energy. So, when electromagnetic radiation is absorbed, the energy it carries will make the absorbing material hotter. It may even produce an alternating current (a.c.) with the same frequency as the radiation in the absorbing material.

Figure 11.10 The energy carried by infra-red radiation soon warms us up

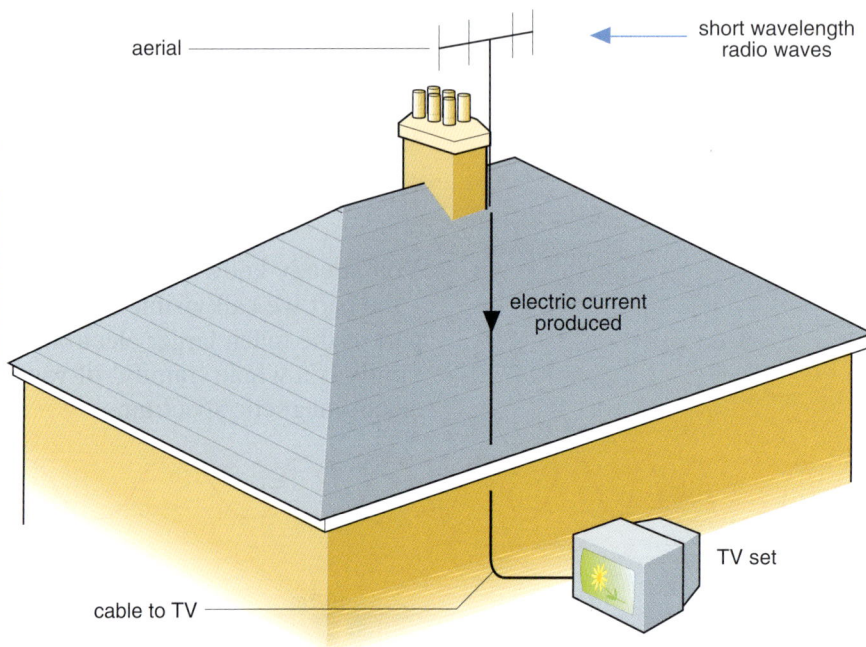

short wavelength radio waves

aerial

electric current produced

cable to TV

TV set

Figure 11.11 Electromagnetic radiation can produce an alternating current

11.3 Using microwaves, infra-red and ultraviolet waves in the home

Microwave cooking

So how does a microwave oven cook food? The answer might be obvious. You put the food in, push a few buttons and the food becomes hot. But of course it's not as simple as that. In fact scientists have more than one way of explaining how microwaves cook food. The neatest explanation is that microwaves which can pass easily through materials such as plastic, paper and ceramic are strongly absorbed by water molecules. The energy carried by the microwaves heats up the water. As most foods contain a lot of water, this gets heated by the microwaves and in turn cooks the food.

Food cooked in a conventional gas or electric oven is heated by conduction. This is a relatively slow process. Microwaves heat the food more quickly by penetrating several centimetres below the surface before being absorbed. So, using a microwave oven can reduce cooking times by about a quarter.

7 Explain why microwaves defrost frozen food rapidly.

8 Microwaves of frequency 2450 MHz are used to heat food. Calculate the wavelength of these microwaves.

9 Why is it important that microwaves don't leak from a microwave oven?

Activity – Sharon investigates microwaves

Sharon noticed that all the food containers she uses in her microwave are made from plastic. She thinks this is because microwaves lose less energy when they pass through a plastic container than they do through a ceramic container.

To test her idea, Sharon measured $250\,cm^3$ of water into a plastic beaker. She placed the beaker in a microwave oven and heated it on full power for 2 minutes. Every 20 seconds Sharon stopped the microwave and took out the beaker. She stirred the water and took its temperature. Sharon then repeated the experiment heating the same volume of water, but this time in a ceramic coffee mug.

Table 11.1 shows the results which Sharon obtained when she used the plastic beaker.

Time in seconds	0	20	40	60	80	100	120
Temperature in °C	22	30	40	47	52	66	74

Table 11.1 Sharon's results with the plastic beaker

1. Draw a graph of temperature (vertically) against time (horizontally) for Sharon's results.
2. Sharon has one anomalous result. Which result is this? (Remember an anomalous result is one that does not fit the general pattern.)
3. Why did Sharon stir the water before taking the temperature?
4. What was the temperature of the water after 50 seconds?
5. Estimate how long it would have taken for the water to boil.
6. Why is your answer to question 4 more reliable than your answer to question 5?
7. In Sharon's investigation, which quantity was the dependent variable?
8. Name one control variable which Sharon kept constant during the investigation.
9. The smallest scale division on the thermometer Sharon used was 1 °C. If Sharon had used a thermometer in which the smallest scale division was 0.1 °C would her measurements have been more accurate, more precise or more reliable?
10. Why did Sharon take a set of readings rather than just one at the start and one after 2 minutes?
11. Sharon thinks that microwaves lose less energy when they pass through a plastic container than they do through a ceramic container. If she is right, would her results using the coffee mug show a larger or smaller increase in water temperature? Explain your answer.

Infra-red radiation for cooking and detection

All objects emit and absorb infra-red radiation. The higher the temperature of an object, the more infra-red radiation it emits. Objects that absorb infra-red radiation become hotter.

We cannot see infra-red radiation. But an infra-red camera can be used to 'see' warm objects, even when we can't see them in the dark. The camera detects the different wavelengths of infra-red radiation emitted by a person, an animal or an object, and changes this into visible light that we can see. This type of camera is used by the Army and Police to spot people in the dark, and by fire crews to find people trapped in smoke-filled buildings.

Ultraviolet (UV) radiation – sunbathing and energy-efficient lamps

Figure 11.12 The infra-red radiation emitted by the heating element has been absorbed by the bread

UV radiation is emitted by any object at a sufficiently high temperature. In fact, objects emitting UV radiation must be white hot. The Sun emits

Figure 11.13 Sunbathing can be very pleasant, but you must take care to limit your exposure to the Sun

Figure 11.14 The fluorescent dial of this watch glows in the dark

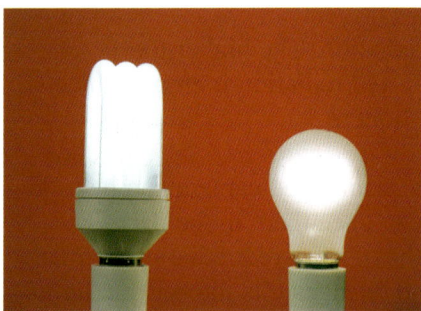

Figure 11.15 Fluorescent lamps are more efficient than ordinary lamps because they do not rely on a metal coil being heated

UV radiation along with infra-red and visible light. UV radiation passes through some substances, but is partially or completely absorbed by others. Most types of glass are very good absorbers of UV radiation.

Some surfaces are good reflectors of UV radiation. If you go skiing it is important to remember that snow reflects up to 90% of the UV that hits it. As UV can damage the retina of your eyes, it's vital to wear glasses or goggles that give the right protection. Without these, you will end up developing snow blindness.

UV radiation given out by the Sun is partly absorbed by ozone in the Earth's upper atmosphere. The UV that passes through the atmosphere is what gives you a sun tan. When you get a tan, one type of skin cell is reacting to the UV, producing a brown pigment called **melanin**. Your skin changes colour and you have a sun tan. Your body produces melanin as its natural defence against UV. By absorbing UV, the melanin protects other cells from damage. But beware – melanin is not produced instantly, it takes time.

It is dangerous to overexpose yourself to UV radiation by being in the Sun for too long. It may cause your skin to age prematurely and increase your risk of getting skin cancer. Listening to weather reports, which include information about the UV level (the 'sun-index'), will help you to limit your exposure. But what else can you do? Here are a few suggestions:
- wear a hat with a wide brim;
- wear Sun-protective clothing. These garments will have a label giving the UPF (Ultraviolet Protection Factor);
- stay out of the Sun between 10a.m. and 3p.m.;
- slap on lots of high SPF (Sun Protection Factor) cream.

It's not all bad news though. UV enables your body to produce vitamin D, which is essential for healthy growth.

Energy-efficient lamps

Some substances can absorb the energy from ultraviolet radiation and then emit the energy as visible light. This is called **fluorescence**. Fluorescent paints and dyes, which seem to glow, work like this.

Fluorescent lamps work by producing ultraviolet radiation. When electricity passes through the gas inside a fluorescent lamp, reactions occur and ultraviolet radiation is emitted. The ultraviolet radiation is absorbed by a chemical which covers the inside of the glass. The chemical fluoresces and visible light is emitted.

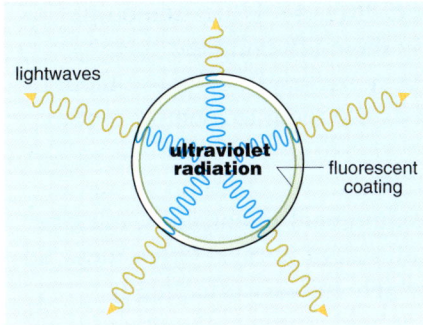

Figure 11.16 Ultraviolet radiation is absorbed and visible light emitted by the fluorescent coating inside the energy-efficient lamps

⑩ Explain why it is difficult to get a sun tan inside a glass conservatory.

⑪ Why are fluorescent lamps often described as 'energy-efficient'?

Figure 11.17

Activity – Sensible sunbathing

A recent report has criticised travel agents for not doing enough to warn their customers of the dangers of sunbathing. Despite a sharp rise in the number of deaths from skin cancer, travel brochures still show lots of photos of scantily dressed, sun-tanned people lying around in the Sun.

Write a short article for the front page of a travel brochure. The article must warn people about the risk of skin cancer and, most importantly, give advice on how to reduce this risk. Of course, the article must not deter potential customers from taking a holiday in the Sun.

11.4 Using X-rays and gamma rays in medicine

X-rays and gamma rays are at the short wavelength end of the electromagnetic spectrum. There is no sharp dividing line between the two. The range of wavelengths of X-rays overlaps that of gamma rays. Both types of radiation can penetrate and pass through some solid materials with very little energy being absorbed. Just like visible light, they also affect photographic film.

Both X-rays and gamma rays are **ionising** radiations. This means they can change the atoms of the materials they pass through. This is why they are harmful to us. When X-rays and gamma rays are absorbed by body tissues, their energy can remove electrons from some atoms, forming positive ions. This process is called **ionisation.** It can change the molecules that control the way in which a cell operates. This may result in damage to the central nervous system, mutation of genes or even cancer. So exposure to ionising radiations like X-rays and gamma rays must always be kept to a minimum.

X-ray photography

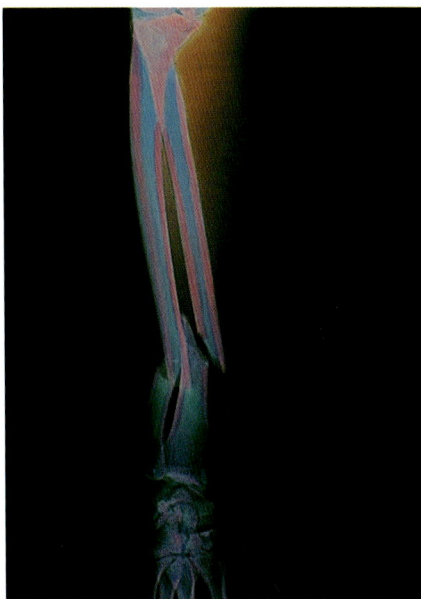

Figure 11.18 An X-ray photograph of a broken arm

X-rays are absorbed by all body tissues to some extent. Bones, teeth and diseased tissues absorb more energy than healthy tissues, which X-rays pass through very easily.

In an X-ray photograph, bones, teeth and diseased tissues stand out because they absorb the X-rays and stop them affecting the photographic film. This makes X-rays very useful. Doctors use them to show where a bone is broken and dentists use them to identify tooth decay.

Radiographers, who operate the X-ray equipment in hospitals, must be protected from the dangerous ionising effects of X-rays. Often they work behind lead or concrete screens which are very effective in absorbing X-rays. If they cannot work behind a protective screen, then they wear a lead-lined apron. Lead is also used to protect the parts of a patient's body that are not being X-rayed.

Gamma rays

Gamma rays are widely used in diagnosing and treating diseases such as cancer. This is taken up and covered fully in Section 11.12.

12 Why is X-ray film kept in light-proof containers?

13 What precautions should a dentist take before X-raying a patient's tooth?

14 Why are X-rays and gamma rays described as ionising radiations?

Career – Radiographer

What does a radiographer do? Radiographers are involved in both the diagnosis and treatment of illness. A diagnostic radiographer uses a variety of techniques to produce images of our bodies. Using X-ray machines is just one part of the job. A therapy radiographer is an important part of the team that treats a patient. The radiographer will be involved in planning and delivering the treatment using high-energy X-rays and gamma rays. Working as part of a team, radiographers also need good communication skills. If you want to find out more about a career as a radiographer go to www.radiographycareers.co.uk or www.sor.org.

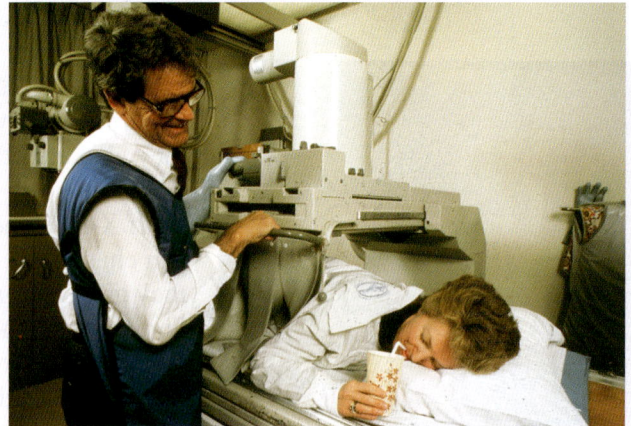

Figure 11.19 This radiographer, wearing a protective lead-lined apron, is preparing a patient for an X-ray

11.5 Using waves in communications

Visible light, infra-red, microwaves and radio waves are essential for our communications.

Visible light and infra-red

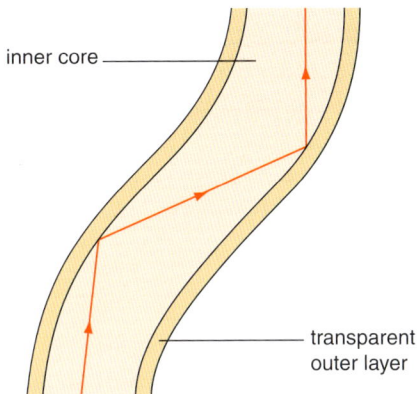

inner core

transparent outer layer

Figure 11.20 Reflection allows visible light and infra-red to pass along the optical fibre

Thin rods of glass, called **optical fibres**, are used to carry both infra-red and visible light rays. The fibres, no thicker than a human hair, consist of two parts – an inner core through which the light or infra-red travels and an outer layer that protects the inner core from being scratched. Once an infra-red or visible light ray is inside the fibre, it is reflected every time it hits the boundary between the inner core and outer layer. In this way, the ray travels from one end of the fibre to the other. It even follows the bends and twists in the fibre.

For communications, digital signals are produced by converting speech and electronic messages to light or infra-red impulses (see Section 11.6). These signals can then be transmitted along the optical fibres. Many telephone links now use optical fibres rather than copper cables. A single fibre is capable of carrying thousands of telephone conversations at the same time.

Microwaves and radio waves

The microwave and radio wave sections of the electromagnetic spectrum are divided into a number of wavelength bands. Together these bands cover a huge range of wavelengths, from 1 mm to over 100 km. The properties of the waves within each band determine the way in which the waves are used for communications.

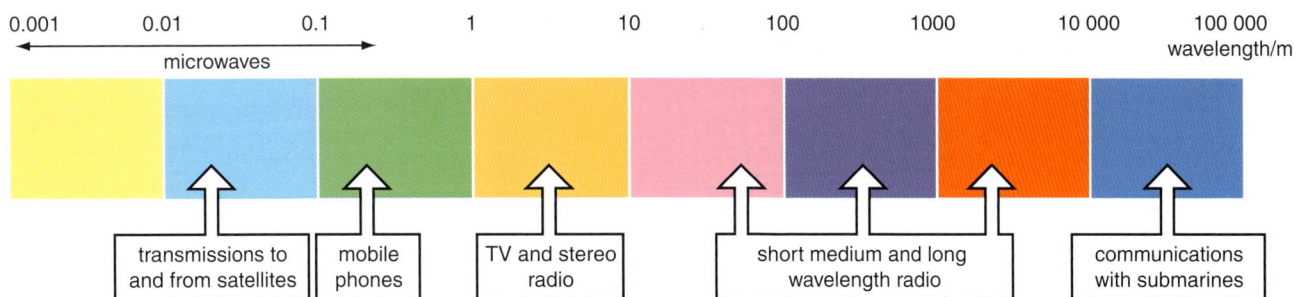

| 0.001 | 0.01 | 0.1 | 1 | 10 | 100 | 1000 | 10 000 | 100 000 wavelength/m |

microwaves

transmissions to and from satellites

mobile phones

TV and stereo radio

short medium and long wavelength radio

communications with submarines

Figure 11.21 Each band of the microwave and radio wave spectrum has a specific communications use

Radio waves in the short, medium and long wavelength bands are transmitted around the Earth by reflections from the **ionosphere** (layers of ionised gas in the Earth's upper atmosphere). Radio stations, which use radio waves with wavelengths of 10 m or more, can therefore broadcast over very long distances with the shortest wavelengths travelling worldwide.

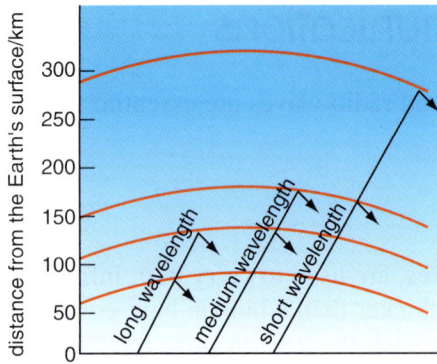

Figure 11.22 Different wavelength radio waves are reflected by different layers of the ionosphere

More information can, however, be transmitted by shorter wavelength, higher frequency waves. So, stereo radio is broadcast using very high frequency (VHF) waves and television uses ultra high frequency (UHF) waves. Unfortunately, these waves are not reflected by the ionosphere, so their range is limited by the Earth's curvature. National broadcasts, using narrow beams of these waves, are made from communication towers such as the British Telecom Tower in London. These narrow beams are passed through a network of repeater stations throughout the country. At each station, the signal is transmitted in all directions to our homes.

Microwaves, with wavelengths of only a few centimetres, pass easily through the Earth's atmosphere. So, information is carried by microwaves to and from satellites. Many people now receive their television programmes from signals carried by microwaves. These have travelled from a transmitter on Earth to a satellite and then back to a dish which acts as an aerial on the side of their house.

Nowadays, the commonest use of microwaves is in mobile phone networks. These microwaves are transmitted from place to place via tall aerial masts or from one continent to another via satellites.

Figure 11.23 Stereo radio and TV programmes are transmitted from telecommunications towers, such as the BT Tower in London

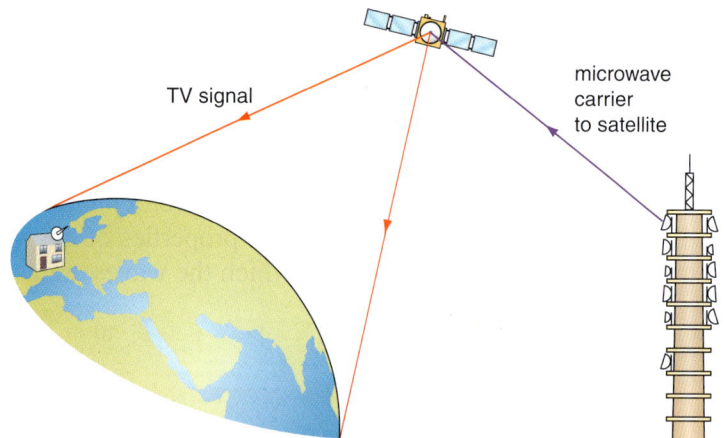

Figure 11.24 A satellite can transmit a microwave signal that covers a large part of the Earth's surface

---

⑮ Explain how an optical fibre is able to carry a telephone conversation.

⑯ A TV programme in the UK uses a live satellite link to question a reporter in the USA. Why is there a delay between the question being asked and the reporter hearing it?

⑰ Why do most radio stations broadcast using

waves with wavelengths between 100 m and 10 000 m?

⑱ What property of microwaves allows them to transmit information to satellites?

⑲ Which radio band can be used to communicate with a nuclear submarine on patrol in the North Sea?

Activity – Are mobile phones a health risk?

At present the simple answer to this question is that nobody really knows. But there are over 40 million mobile phones in the UK alone, so many people would like an answer.

In order to understand why mobile phones may present a health risk, we need to know how they work. Mobile phones are designed to transmit microwaves from an aerial in all directions. They must do this because the mast that receives and sends on the signal could be in any direction from where the phone is used. This means that some of the microwaves will travel towards us and penetrate our bodies, particularly our heads. As we learnt in Section 11.3, microwaves are strongly absorbed by water and our bodies contain a lot of water. The energy absorbed by the water will raise the temperature of body tissues and may damage the cells (Figure 11.25).

Several scientific studies have suggested that there could be a link between mobile phone use and health problems such as tiredness, headaches, memory loss and brain tumours. Although the evidence from these studies is *not* clear-cut and conclusive, it indicates a need for more research.

In a report produced in 2005, Professor Sir William Stewart of the National Radiological Protection Board said,

'I don't think we can put our hands on our hearts and say mobile phones are safe. If there are risks – and we think there may be risks – the people who are going to be most affected are children, and the younger the child, the greater the danger.'

Following the Stewart report, a mobile phone specifically designed for children under the age of eight was withdrawn from sale in the UK. Some

Figure 11.25 Some energy from the microwaves is absorbed by the head and body

people did not agree with this decision. Many parents want their children to have a mobile phone for emergencies. They argue that they worry less if their children can contact them at any time.

❶ What does a scientist mean by conclusive evidence?

❷ What reasons could a manufacturer give for producing a mobile phone for young children?

❸ Why do you think a young child is more likely to be affected by the radiation from a mobile phone than an adult?

❹ A study of 750 people in Sweden led to the suggestion that using a mobile phone for at least 10 years increases the risk of a tumour on the nerve between the ear and the brain by four times. The study has not been repeated.

  a) Do you think the number of people studied was sufficient to give firm evidence of a link between mobile phone use and developing a tumour? Give reasons for your answer.

  b) Why would it be a good idea to repeat the study?

  c) Why was the study restricted to people who had used a mobile phone for at least 10 years?

  d) If there was clear scientific evidence that using a mobile phone increased the risk of developing a tumour by four times, would you still use one? Give reasons for your answer.

Why are digital signals taking the place of analogue signals?

People have always wanted to communicate and nowadays it's easier than ever. Just pick up the phone, use your mobile or send an e-mail. We can communicate with people almost anywhere in the world and it is virtually instantaneous. But it hasn't always been this easy.

The telegraph was the first modern invention that allowed information to be sent using electricity as a message carrier. The signal was sent using a series of current pulses. These pulses were the first form of **digital signals**. In the USA telegraphy quickly became the fastest way of communicating over thousands of kilometres. But in modern terms, telegraphy was slow. At one end, an operator needed to code the message and then send the signal. At the other end, another operator wrote down the signal and then decoded it.

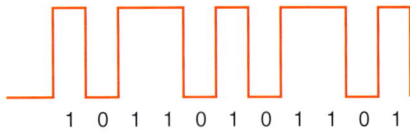

Figure 11.26 A digital signal is either 'ON' or 'OFF'

1 0 1 1 0 1 0 1 1 0 1

In modern digital communication systems, complex electronic circuits have replaced the operators. These circuits produce signals rapidly using only two electrical values, a high voltage pulse (ON) and a low voltage pulse (OFF). The number **1** is used to represent the 'ON' state and the number '**0**' the 'OFF' state. This allows any digital signal to be written as a series of ones and zeros.

The telephone, invented in 1876, was a major breakthrough in electrical communication. For the first time people were able to talk directly to each other. When you speak into a telephone, you produce sound waves. A microphone inside the telephone changes these sound waves into electrical signals. The amplitude and frequency of the electrical signals change as the amplitude and frequency of the sound waves change. These signals are called **analogue signals**. Analogue signals are waves with amplitudes and frequencies that vary continuously from zero upwards. In contrast, digital signals are integral values using the numbers 1 and 0.

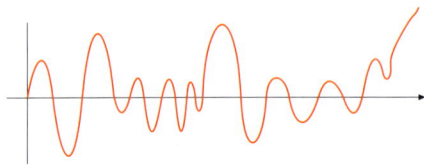

Figure 11.27 An analogue signal produced by a microphone

In older telephone systems, copper cables carry the analogue signals to a receiver. The earphone in the receiver acts like a small loudspeaker. It changes the electrical signals back into sound. In modern telephone systems optical fibres, microwave links and radio links are replacing the copper cables linking 'sender' to 'receiver'. (See also Section 11.5.)

So communication signals may be either digital or analogue. But digital signals have certain advantages over analogue signals.

- Two things happen when any signal travels from a transmitter to a receiver. Firstly, the signal will get weaker. Secondly, it will pick up unwanted signals called 'noise' that **interfere** with and distort the original signal. With an analogue signal, different frequencies weaken by different amounts. So, during any amplification process to make the weak signal stronger, the original signal gets more and more distorted. With a digital signal, any unwanted 'noise' usually has a

20 Say briefly why digital signals are taking over from analogue signals.

21 Why is a radio often called a 'wireless'?

22 Describe what is meant by a digital signal.

23 Explain why the quality of a digital signal transmitted over a long distance does not change.

low amplitude. The receiver still recognises the original 'on' and 'off' states. This means that the digital signals received are not distorted and have a quality that matches the original signal.
- Computers work and communicate using digital signals. This makes it easy for a computer to process data in the form of a digital signal.

What are atoms really like?

11.7

Figure 11.28 J.J. Thomson (1856–1940) discovered electrons in 1897 when he was Professor of Physics at Cambridge University. In 1906, Thomson was awarded the Nobel Prize for Physics. Thomson was a brilliant teacher – seven of his students won Nobel Prizes for Physics or Chemistry.

Just over a century ago, scientists thought that atoms were hard, solid particles like tiny invisible marbles. Then, in 1897, J.J. Thomson discovered that the outer parts of atoms contained tiny negative particles, which he called **electrons**. As atoms were neutral overall, Thomson's discovery of negative electrons led scientists to think that the central part of an atom (the **nucleus**) must be positive. In 1909, Ernest Rutherford found evidence for positive particles in the nucleus and called them **protons**. Further experiments suggested that the nuclei of atoms must also contain neutral particles as well as protons. In 1932, James Chadwick, working with Rutherford, discovered these neutral particles and called them **neutrons**.

We now know that:
- all atoms are made up from three basic particles – protons, neutrons and electrons;
- the nuclei of atoms contain protons and neutrons;
- protons and neutrons have the same relative mass of one;
- protons have a positive charge, but neutrons are neutral;
- more than 99% of an atom is empty space occupied by negative electrons;
- the mass of an electron is about 2000 times less than that of a proton;
- the negative charge on one electron just cancels the positive charge on one proton;
- electrons whiz around the nucleus very rapidly. They occupy layers, or shells, at different distances from the nucleus.

The key points about atomic structure are summarised in Table 11.2.

Figure 11.29 Ernest Rutherford (1871–1937) found evidence for protons in 1909. In 1895, Rutherford left New Zealand to work with J.J. Thomson at Cambridge University. Later, he moved to Manchester University where he won the Nobel Prize for Chemistry in 1908. In 1919, Rutherford succeeded J.J. Thomson as Professor of Physics at Cambridge. Rutherford was a brilliant experimental scientist. His experiments always seemed to work.

❷❹ Copy out Table 11.2 and fill in the blank spaces.

Particle	Position in the atom	Relative mass (atomic mass units)	Relative charge
Proton	....	1	....
Neutron	Nucleus	....	0
....	Shells	$\frac{1}{2000}$	−1

Table 11.2 The positions, relative masses and relative charges of protons, neutrons and electrons

Protons, neutrons and electrons are the building blocks for all atoms. Hydrogen atoms are the simplest, with just one proton and one electron (Figure 11.30a). The next simplest atoms are those of helium with two protons, two electrons and two neutrons. Next comes lithium with three protons, three electrons and four neutrons (Figure 11.30b).

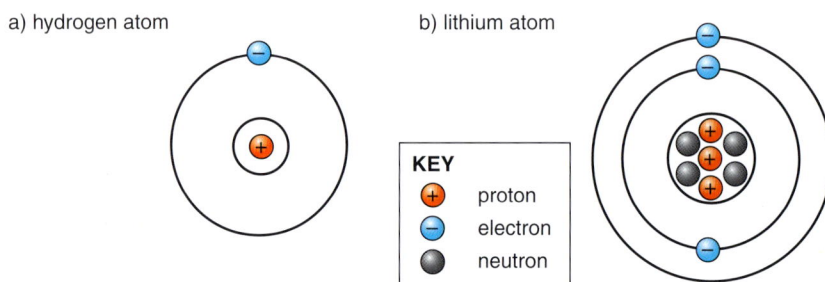

Figure 11.30 Protons, neutrons and electrons in a hydrogen atom and a lithium atom

Some of the heaviest atoms have large numbers of protons, neutrons and electrons. For example, gold atoms have 79 protons, 79 electrons and 118 neutrons.

Notice, in all these examples, that **an atom always has the same number of protons and electrons**. In this way, the positive charges on the protons just cancel the negative charges on the electrons and the atoms have no overall charge.

Atomic number and mass number

The only atoms with one proton are those of hydrogen. The only atoms with two protons are those of helium and the only atoms with 79 protons are those of gold.

So, the number of protons in an atom tells you which element it is. The number of protons in an atom is called its **atomic number**.

Hydrogen atoms have one proton. So, hydrogen has an atomic number of one, helium has an atomic number of two and gold has an atomic number of 79.

The mass of the electrons in an atom can be ignored compared with the mass of protons and neutrons. This means that the mass of an atom depends on the number of its protons plus neutrons. And the total number of protons plus neutrons in an atom is called its **mass number**.

**So, atomic number = number of protons
and mass number = number of protons + number of neutrons.**

Figure 11.31 If the nucleus of an atom is enlarged to the size of a pea and put on the top of Nelson's Column, the outermost electrons will be as far away as the pavement below.

25 Lithium atoms have three protons and four neutrons.
a) What is the atomic number of lithium?
b) What is the mass number of lithium?
c) How many electrons are there in a lithium atom?

Sometimes, the symbol Z is used for atomic number and the symbol A for mass number. So, for lithium, Z = 3 and A = 7. Figure 11.32 shows how the mass number and atomic number are often shown with the symbol of an element.

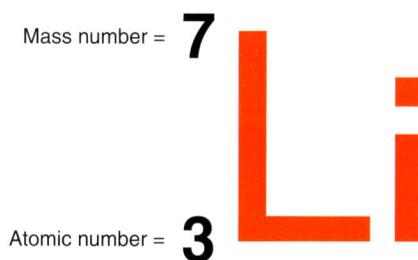

All the atoms of one element have the same number of protons, the same number of electrons and therefore the same atomic number. But, the atoms of one element can have different numbers of neutrons and therefore different mass numbers. These atoms of the same element with different mass numbers are called **isotopes**.

Mass number = **7**

Li

Atomic number = **3**

Figure 11.32 The mass number and atomic number shown with the symbol for lithium (Li)

For example, naturally occurring chlorine has two isotopes, $^{35}_{17}Cl$ called chlorine–35 and $^{37}_{17}Cl$ called chlorine–37. Each of these isotopes has 17 protons, 17 electrons and the same atomic number. They also have the *same chemical properties* because these are determined by the number of electrons. However, one isotope ($^{35}_{17}Cl$) has 18 neutrons and the other ($^{37}_{17}Cl$) has 20 neutrons. Therefore, they have different mass numbers and *different physical properties* because these depend on the masses of the atoms.

The similarities and differences between isotopes of the same element are summarised in Table 11.3.

Isotopes have the same	Isotopes have different
• Number of protons	• Numbers of neutrons
• Number of electrons	
• Atomic number	• Mass numbers
• Chemical properties	• Physical properties

Table 11.3 The similarities and differences between isotopes

26 Copy and complete the table below for the isotopes of carbon.

	Carbon–12, $^{12}_{6}C$	Carbon–14, $^{14}_{6}C$
number of protons	6	
number of electrons		
number of neutrons		
mass number		

11.8 Radioactive materials

Most radioactive materials that we use today have been synthesised by chemists, but some occur naturally. Large amounts of radioactive uranium and plutonium are used to generate electricity in nuclear power stations. Tiny amounts of these and other radioactive substances can be

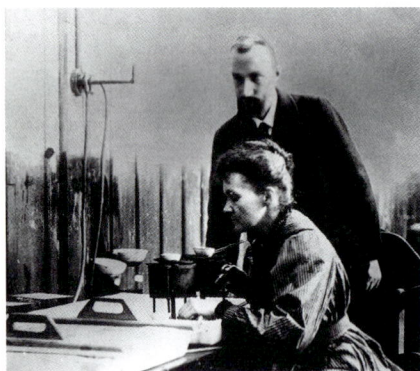

11.9

used to generate electricity in heart pacemakers. Radioactive materials can be beneficial, but they can also be harmful. They can cause cancer and also cure it.

A Frenchman, Henri Becquerel, and his Polish assistant, Marie Curie, carried out the first investigations into radioactivity in 1896. They discovered that all uranium compounds emitted **radiation**. This radiation could pass through paper and affect photographic film like light. Becquerel called the uranium compounds **radioactive substances** and he described the process by which they emit radiation as **radioactivity** or **radioactive decay**.

What is the radiation from radioactive substances?

The best way to detect the radiation from radioactive materials is to use a Geiger-Müller tube (Figure 11.34). When radiation enters the tube, atoms inside are ionised (changed into ions). These ions are attracted to electrodes in the tube and a small current flows. This current is then amplified and a counter is used to count the amount of radiation entering the tube. The amount of radiation is usually measured in counts per second or counts per minute.

Experiments show that radioactive atoms give out radiation all the time, whatever is done to them. The rate at which they decay is not affected by changes in temperature or by different atoms combined with the radioactive atoms. The radiation emitted consists of tiny particles and rays of electromagnetic waves which come from the nuclei of the radioactive atoms.

The nuclei of radioactive atoms emit three kinds of radiation – **alpha particles** (α particles), **beta particles** (β particles) and **gamma rays** (γ rays).

When radioactive atoms emit alpha particles or beta particles they become different atoms with different numbers of protons and neutrons in the nucleus. So, the emission of alpha and beta radiation leads to changes in the nuclei at the centre of radioactive atoms. This is different from chemical reactions, which involve changes in the electrons in the outer parts of atoms (see Section 5.2). When radioactive atoms emit gamma rays there is no change in the number of protons or neutrons in the nucleus but the atoms do lose energy. The energy is lost as penetrating electromagnetic radiation, similar to light.

The characteristics and properties of the three kinds of radiation are summarised in Table 11.4.

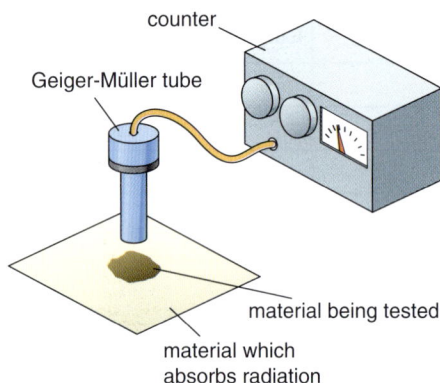

Figure 11.34 A Geiger-Müller tube and counter detecting radiation from a radioactive substance

Radiation	Nature of radiation	Penetrating power	Effect of electric and magnetic fields	Ionising power
Alpha particles	Helium nuclei containing two protons and two neutrons, $^4\text{He}^{2+}$	Travel a few centimetres through air, absorbed by thin paper	Very small deflection	Strong
Beta particles	Electrons, e^-	Travel a few metres through air, pass through paper, but absorbed by 3 mm of aluminium foil	Large deflection	Moderate
Gamma rays	Electromagnetic waves	Travel a few kilometres through air, pass through paper and aluminium foil, but absorbed by very thick lead	None	Weak

Table 11.4 The characteristics and properties of alpha, beta and gamma radiations

27 A scientist divided 3.0 g of pure uranium into three 1.0 g samples. He then converted these samples, without any loss of uranium, into separate samples of uranium oxide, uranium chloride and uranium bromide. He then measured the radioactivity of the three samples at the same time and temperature.
a) Are the three uranium compounds a categoric variable, a discrete variable or a continuous variable?
b) What is the dependent variable in this investigation?
c) Will the results for the three samples be:
  i) exactly the same;
  ii) more or less the same;
  iii) very different?
d) Explain your answer to part c).

Notice the different penetrating power of the three radiations in Table 11.4. These differences in penetrating power are emphasised in Figure 11.35. Gamma rays are much more penetrating than beta particles and beta particles are much more penetrating than alpha particles:
- alpha particles cannot even pass through a sheet of paper;
- beta particles can pass through a sheet of paper but are absorbed by 3 mm aluminium foil;
- gamma rays can pass through a sheet of paper and 3 mm aluminium foil but are absorbed by thick lead. Because of this, radioactive substances are usually stored in thick lead containers.

Although alpha particles are the least penetrating, they have the strongest ionising power. On the other hand, gamma rays are the most penetrating but they have the weakest ionising power.

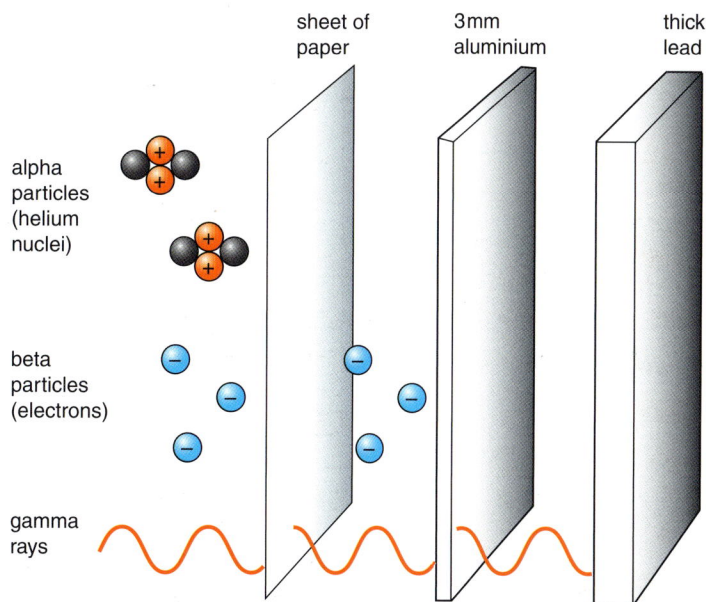

Figure 11.35 The relative penetrating power of alpha particles, beta particles and gamma rays

28 Figure 11.36 shows a thin beam of alpha particles being deflected by an electric field.
a) Why are the alpha particles deflected towards the negative plate?
b) Copy the diagram and assume that the source of radiation also emits beta particles. Draw the path of beta particles on your diagram. Remember that beta particles are negatively charged electrons and these are much lighter than alpha particles.
c) Now assume that the source of radiation also emits gamma rays. Draw the path of gamma rays on your diagram.

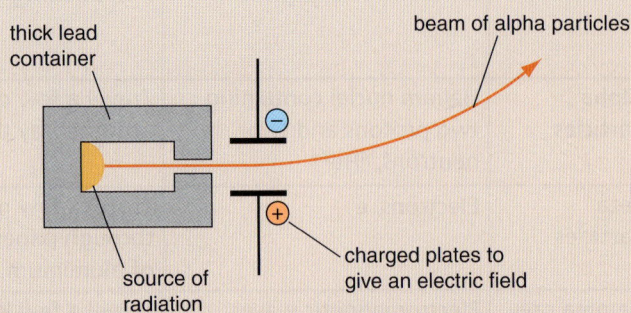

Figure 11.36 The effect of an electric field on alpha radiation

11.10 How much radiation are we exposed to?

Every day we are exposed to radiation from the Sun, from rocks and even from our food. This natural radiation to which we are exposed is called **background radiation**. As background radiation comes from various sources, it is higher in some areas than others. We are exposed to it all our lives. But, normally it is very low and no risk to our health.

Although background radiation is very small, it must be accounted for in accurate measurements of radioactivity. This involves subtracting the average count rate for background radiation from the measured count rate of the sample being tested.

Figure 11.37 This environmental scientist is using a Geiger-Müller tube to check the levels of radioactivity in coastal soils

29 An environmental scientist was asked to check the radiation from various plants along a path. First, she measured the background radiation by pointing her Geiger-Müller tube in different directions at six points along the path. She then measured the radiation level near six different plants along the path by holding her Geiger-Müller tube 20 cm in front of each plant. Her results are shown below.

Background radiation at six points along the path / counts per minute	58, 59, 58, 62, 60, 63
Radiation level in front of six plants / counts per minute	70, 64, 63, 60, 67, 66

a) Why did she measure the background count at six points along the path?
b) What value should she take for the background radiation in counts per minute?
c) What is the average radiation level in front of the six plants?
d) What is the average level of radiation emitted by the plants after correction for background radiation?

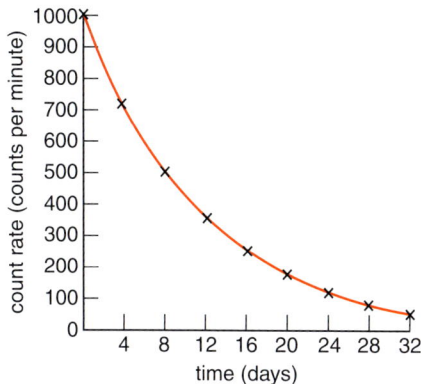

Figure 11.38 A radioactive decay curve for iodine–131

11.11 Radioactive decay

The breakdown of unstable radioactive nuclei is a random process. You cannot tell when an unstable nucleus will break down. But, if there are large numbers of unstable atoms, an average rate of decay will occur. So, if a sample contains 10 million radioactive atoms, it should decay at twice the rate of a sample containing 5 million atoms. And, as the unstable nuclei decay, the rate of decay will fall.

Figure 11.38 shows the decay curve for a sample of iodine–131. Doctors use this isotope to study the uptake of iodine by the thyroid glands. The shape of the decay curve is similar to those for other radioactive materials, but the scales on the axes can vary a great deal.

Notice from Figure 11.38 that:
- the count rate falls from 1000 to 500 counts/minute in 8 days;
- the count rate falls from 500 to 250 counts/minute in 8 days;
- the count rate falls from 720 to 360 counts/minute in 8 days.

These results show that the time taken for the count rate of iodine–131 to fall to half its initial value is constant at eight days. We can also conclude from this that the time taken for the number of iodine–131 atoms to halve is also eight days. A similar constant pattern, in halving the count rate and the number of radioactive atoms, occurs with all radioactive substances. The time it takes for the count rate or for the number of atoms of a radioactive isotope to fall to half its initial value is called its **half-life**.

Half-lives can be determined by plotting graphs of count rates against time, like that in Figure 11.38, and then finding an average time for the count rate to fall by half.

Half-lives can vary from a few milliseconds to millions of years. The shorter the half-life, the faster the isotope decays and the more unstable it is. The longer the half-life, the slower the decay process and the more stable the isotope. Polonium–234 with a half-life of only 0.15 milliseconds is very, very unstable. Uranium–238 with a half-life of 4500 million years is 'almost stable'.

What are the hazards of using nuclear radiation?

The particles and electromagnetic waves in nuclear radiation can cause the atoms in different materials to ionise. When an atom is ionised, one or more electrons are removed from it or added to it. If ionisation occurs in our bodies, the reactions in our cells may change or stop and cause disease. Because of this, exposure to ionising radiations can be harmful.

People who work with radioactive materials must be aware of the dangers from radiation.

30 A hospital gamma ray unit contains 10 grams of cobalt–60. This has a half-life of five years. This means that half the cobalt–60 will have decayed after five years leaving only five grams of cobalt–60.
   a) How much cobalt–60 will be left after another five years (i.e. 10 years from the start)?
   b) How much cobalt–60 will be left after 15 years from the start?

Figure 11.39 This technician is wearing a special radiation-sensitive badge. This contains photographic film to show the level of radiation to which he has been exposed.

- Low doses of radiation cause nausea and sickness.
- Moderate doses of radiation cause damage to the skin and loss of hair.
- High doses of radiation cause sterility, cancers and even death.

Scientists and technicians who work with low-level radioactive materials must wear special badges containing photographic film sensitive to radiation. The film is developed at regular intervals to show the level of exposure.

The effects of radiation depend on its penetration, its ionising power and its half-life, as well as the length of exposure.

In general, gamma rays with their greater penetrating power are more harmful than alpha and beta particles. So, people who work with dangerous isotopes emitting gamma rays must take extra safety precautions. These include:
- shields of lead, concrete or thick glass to absorb the radiation;
- lead aprons, worn by radiographers who work continually with X-rays and gamma rays;
- remote-control handling of dangerous isotopes from a safe distance;
- reducing exposure to radiation to the shortest possible time.

Activity – What are the long-term effects of radiation?

The data in Table 11.5 show the effects of radiation on uranium miners in Russia and on the people of Hiroshima after the atomic bomb in 1945.

❶ There were 100 extra deaths due to cancer in the 15 000 people living in Hiroshima after the atomic bomb had been dropped. This works out at one extra death per 150 people.
  a) How many people relate to one extra death with the uranium miners?
  b) Did the people living in Hiroshima or the uranium miners in Russia suffer the most?

  c) Do the data in Table 11.5 suggest that alpha radiation is more dangerous or less dangerous than gamma radiation?

❷ a) What was the source of radiation which affected the uranium miners?
  b) What form or forms of cancer do you think the miners suffered from? Explain your answer.

❸ It would be wrong to jump to conclusions about the data in Table 11.5 because it was not obtained by a fair test. The comparison of Russian miners with the people in Hiroshima was not carried out in a fair way.
  a) Explain what is meant by a fair test.
  b) State six reasons why it is wrong to compare the data for Russian miners with those for the people of Hiroshima.

Source of radiation	Type of radiation	Number of people studied	Extra deaths due to cancer caused by the radiation
Radon gas from decay of uranium	Alpha particles	3400 uranium miners in Russia	60
Radioactive materials from the Hiroshima bomb	Gamma rays	15 000 inhabitants of Hiroshima	100

Table 11.5 The effects of radiation on uranium miners and the people of Hiroshima

11.12 What are the uses of radioactive materials?

Radioactive isotopes are widely used in medicine and in industry. Their uses depend mainly on their penetrating power and half-life.

Medical uses

Although radiation can damage our cells, it can also be used to treat cancers. Cancer cells are damaged and killed more easily by radiation than healthy cells. This is because cancer cells grow and divide more rapidly than healthy cells.

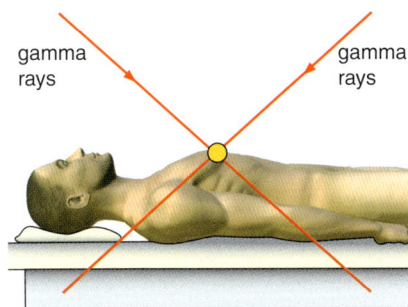

gamma rays

gamma rays

Figure 11.40 Treating lung cancer using gamma rays

Look at Figure 11.40. This shows how radiotherapy is used to treat a cancerous growth. The process requires penetrating gamma rays, which can pass through flesh and kill cancerous cells inside the body. It also requires an isotope with a fairly long half-life so that the gamma radiation can be adjusted to the same level over a few weeks of treatment. Cobalt–60 ($^{60}_{27}$Co), which emits gamma rays with a half-life of 5.3 years, is widely used.

Figure 11.40 shows beams of gamma rays from cobalt–60 being used to treat lung cancer. Gamma rays hit the cancer from several directions. So, the cancer gets a much higher dose than the surrounding lung tissue.

Skin cancer can be treated with less penetrating beta rays. This is done by strapping a plastic sheet containing phosphorus–32 ($^{32}_{15}$P) with a half-life of 14 days on the affected area.

Radioactive substances are easy to detect. So, they can be used in medicine to *trace* what happens to chemicals injected into our bodies. The isotopes used for this are often called **tracers** which help doctors in the diagnosis of an illness. They must emit gamma rays, which can penetrate flesh and bone to be detected outside the body. Ideally, they should also have a short half-life so that the level of radiation in the body soon falls to a safe level.

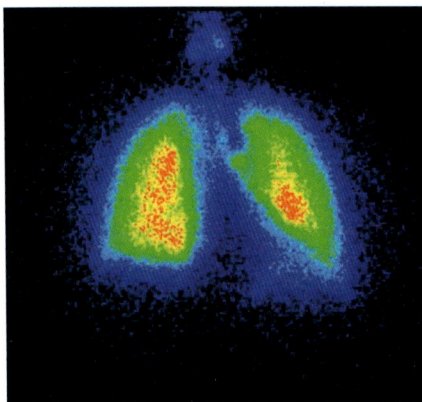

Technetium–99 ($^{99}_{43}$Te) with a half-life of six hours and iodine–131 ($^{131}_{53}$I) with a half-life of eight days (see Section 11.11) are both used as tracers (Figure 11.41).

Most medical equipment, such as dressings and syringes, must be sterilised before being used. This is done by sealing the equipment in plastic bags and exposing them to intense gamma radiation from cobalt–60.

Figure 11.41 After injecting radioactive technetium-99 into a patient's blood, a doctor can study the blood flow through the heart and lungs.

Activity – Choosing radioactive isotopes for different industrial uses

Radioactive isotopes have a large number of industrial uses. These include their use in detecting leaks (Figure 11.42) and in thickness gauges (Figure 11.43). The choice of isotope is based on the type of radiation needed and a suitable half life. This is also the case with medical uses discussed in the last section.

❶ Look carefully at Figure 11.42.
   a) What type of radiation (alpha, beta or gamma) does this use require?
   b) Should the radiation used have a long or a short half-life? (Do you want the radiation to remain in the pipe for a long or a short time?)
   c) Why is the radioactive source described as a tracer when it is used in this way?

❷ Look carefully at Figure 11.43.
   a) What type of radiation (alpha, beta or gamma) does this use require?
   b) Should the radiation used have a long or a short half-life? (Do you want the level of radiation to remain at a constant level for a long time or to disappear fairly quickly?)
   c) What happens to the reading on the Geiger-Müller counter if the sheet of material gets thicker?

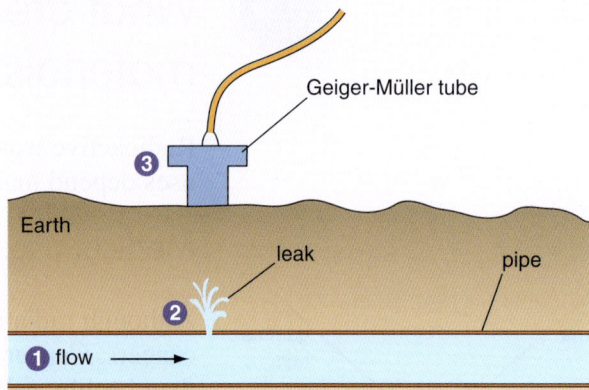

❶ Small amount of radioactive isotope is fed into pipe.
❷ Radioactive isotope leaks into soil.
❸ Geiger-Müller tube detects radiation and position of leak.

Figure 11.42 Using a radioactive source to detect a leak in an underground pipe

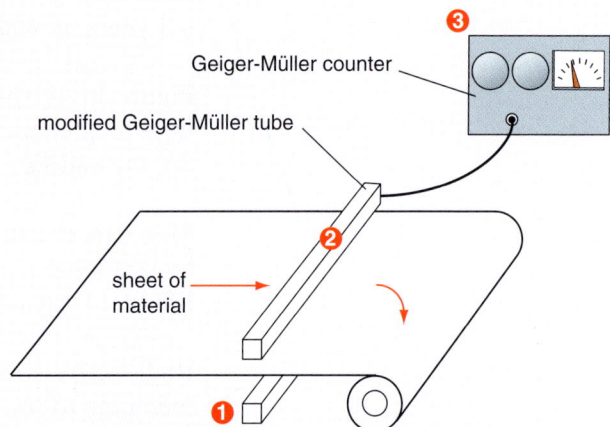

❶ Long radioactive source emits radiation.
❷ Long modified Geiger-Müller tube detects radiation penetrating the thin sheet of paper or plastic.
❸ Geiger-Müller counter measures radiation level. Information is fed back to adjust the thickness of the material if necessary.

Figure 11.43 Using a radioactive source in a thickness gauge for paper or plastic sheets

Summary

✓ All waves move energy from one place to another without transferring any material (matter).

✓ The frequency, wavelength and speed of a wave are linked by the wave equation:

wave speed (m/s) = frequency (Hz) × wavelength (m)

✓ The waves which make up the **electromagnetic spectrum** have several important properties in common.
   • They obey the wave equation.

- They all travel through a vacuum at the same speed of 300 000 000 m/s.
- They can all be reflected, absorbed or transmitted.

✓ When a material absorbs electromagnetic radiation it may get hotter.

✓ When electromagnetic radiation is absorbed, it may produce an alternating current in the absorbing material.

✓ Different parts of the electromagnetic spectrum have different uses.
  - Microwaves and infra-red can be used to cook food.
  - Ultraviolet (UV) radiation causes skin to tan, but it can also cause skin cancer.
  - X-rays and gamma rays have medical and industrial applications.
  - Visible light, infra-red, microwaves and radio waves are used in communications.

✓ **Analogue signals** are continuous waves that vary in amplitude and frequency.

✓ **Digital signals** have only two states, 'ON' and 'OFF'.

✓ Digital signals have two important advantages over analogue signals.
  - Digital signals are less prone to interference than analogue signals.
  - Digital signals can be processed by a computer.

✓ All atoms have a small central nucleus containing protons and neutrons surrounded by layers (shells) of electrons.

✓ All the atoms of one element have the same number of protons.

✓ The number of protons in an atom is called its **atomic number**.

✓ The total number of protons plus neutrons in an atom is called its **mass number**.

✓ Atoms of the same element with different numbers of neutrons and consequently different mass numbers are called **isotopes**.

✓ Some substances give out radiation from the nuclei of their atoms. These substances are said to be **radioactive** and the process is called **radioactivity** or **radioactive decay**.

✓ There are three main types of nuclear radiation emitted by radioactive sources:
  - **alpha particles**, which are helium nuclei, containing two protons and two neutrons with a relative charge of 2+;
  - **beta particles**, which are electrons with a relative charge of 1−;
  - **gamma rays**, which are very penetrating electromagnetic waves.

✓ The natural radiation to which we are exposed all the time is called **background radiation**. Normally, it is very low and there is no risk to our health.

✓ The time it takes for the count rate or for the number of atoms of a radioactive isotope to fall to half its initial value is called its **half-life**.

✓ The effects of nuclear radiation depend on its penetration, its ionising power and its half-life as well as the length of exposure.

✓ Radioactive isotopes have important uses in medicine and in industry. The choice of radiation for a particular use is dictated mainly by its penetration and half-life.

EXAMQUESTION

❶ Table 11.6 gives typical wavelengths for different parts of the electromagnetic spectrum.

Match each wavelength, **A**, **B**, **C** and **D** with the correct part of the electromagnetic spectrum numbered **1**–**4** in Table 11.7.   (*4 marks*)

	Wavelength
A	0.00001 mm
B	0.0006 mm
C	18 cm
D	1.5 km

Table 11.6

Electromagnetic spectrum	
1	Microwaves
2	Ultraviolet
3	Radio
4	Visible light

Table 11.7

❷ a) Explain, with the aid of a diagram, the difference between a digital signal and an analogue signal. (*4 marks*)

b) Why do modern communication systems often transmit digital signals rather than analogue signals? (*2 marks*)

c) Draw a diagram to show how an infra-red signal is carried by an optical fibre. (*2 marks*)

❸ Read the following extract taken from a newspaper article.

Plans for a mobile phone mast in the grounds of a local school have sparked anger amongst parents. Protestors opposed to the mast are worried about the potential health risks and believe that mobile phone masts will eventually be proved harmful. The parents feel that there is an urgent need for unbiased, *reproducible research* to be carried out and until this is done there should be no masts on school grounds. A spokesperson for the telecoms company said 'The mast will be unobtrusive and the health risks are unproven.'

a) Explain why mobile phone masts present a potential health risk. (*3 marks*)

b) What is meant by *reproducible research*? (*2 marks*)

c) Give one reason why the evidence from a research project may be biased. (*1 mark*)

d) There are more than 30 000 mobile phone masts in the UK. Why does this make it difficult for scientists to prove a link between phone masts and health? (*1 mark*)

❹ Several people who work in hospitals and in industry are exposed to different types of radiation. Three ways to check or reduce their exposure to radiation are:
A wear a badge containing photographic film;
B wear a lead apron;
C work behind a thick glass screen with remote-handling equipment.

Which method (A, B or C) should be used by the following people?
a) A hospital radiographer. (*1 mark*)
b) A scientist experimenting with isotopes emitting gamma rays. (*1 mark*)
c) A secretary to a chief radiographer. (*1 mark*)

❺ A Geiger-Müller tube was used to measure a radioactive source with a half-life of 25 years. In five 10-second periods the following number of counts were recorded: 255, 262, 235, 258, 265.
a) Why were the five counts different? (*1 mark*)
b) In order to take fair measurement of the counts, some key variables had to be controlled. Name two variables that should be controlled. (*2 marks*)
c) Why were five counts taken? (*1 mark*)
d) Which one of the five counts is anomalous? (*1 mark*)
e) Calculate the most reliable value for the number of counts per second. (*1 mark*)
f) Suppose five more 10-second counts had been taken. Would this make the result in part e) more precise, more reliable or more valid? (*1 mark*)

❻ Zak's teacher carried out an experiment to measure the half-life of protactinium–234. His results are shown in Table 11.8.
a) In the experiment, what was:
i) the independent variable; (*1 mark*)
ii) the dependent variable? (*1 mark*)
b) What instrument was used to measure the count rate? (*1 mark*)
c) What is the background count during the experiment? (*1 mark*)
d) Re-write the table, showing a corrected count rate at the times shown. (*1 mark*)
e) Plot a graph of count rate (vertically) against time (horizontally). (*3 marks*)
f) How long does it take for the count rate to fall from: i) 60 to 30; ii) 40 to 20; iii) 30 to 15? (*3 marks*)
g) Calculate an average value for the half-life of protactinium–234. (*1 mark*)

Time/s	Count rate/counts/s
0	66
40	44
80	30
120	20
160	13
200	9
240	6
280	4
320	4
360	4

Table 11.8

Chapter 12
What are the origins of the Universe and how is it changing?

At the end of this chapter you should:

✓ know that careful observations can stimulate further investigations and lead to important technological developments;

✓ understand that the results and information from scientific investigations can be made more reliable by repeated observations and the use of more sensitive equipment;

✓ appreciate the advantages and disadvantages of different types of telescopes;

✓ understand how the wavelength and frequency of light and other wave motions change when the wave source is moving;

✓ know that there is a red-shift in light from distant galaxies;

✓ understand how the observed red-shift provides evidence that the Universe is expanding and supports the 'big bang' theory.

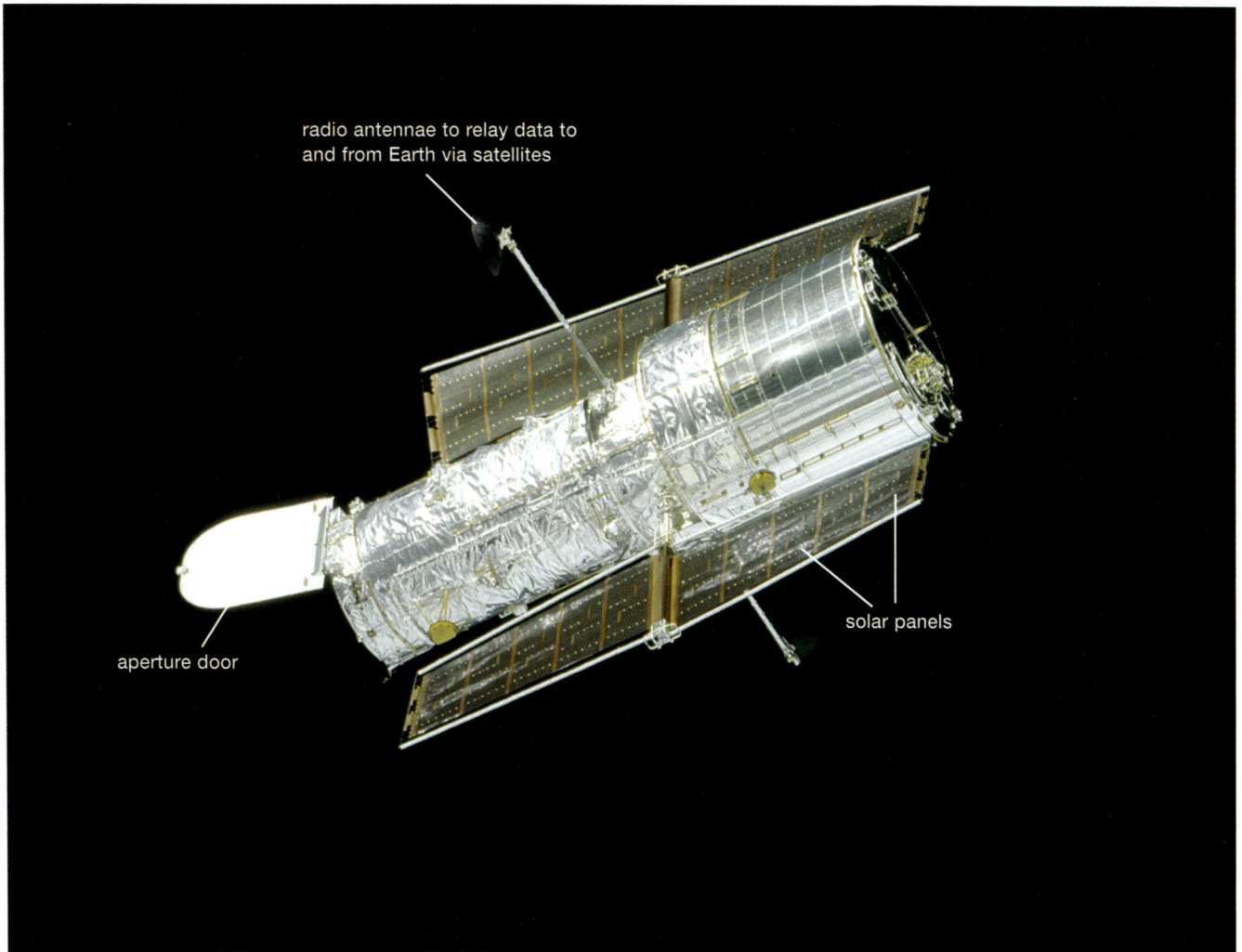

radio antennae to relay data to and from Earth via satellites

solar panels

aperture door

Figure 12.1 The Hubble Space Telescope or HST for short, which orbits the Earth, was launched in April 1990. HST enables astronomers to obtain clearer pictures than ever of the Milky Way and distant galaxies. HST is about the size of a bus. It was designed for launch and repair by astronauts from the Space Shuttle. They can replace worn out parts on the telescope, fit more advanced instruments and re-adjust its orbit.

Stars and galaxies

Betelgeuse, an old red supergiant star.

Rigel, a young bluish supergiant star.

Figure 12.2 Stars that make up the constellation Orion. The black outline shows the picture that the constellation is supposed to represent.

People have been looking at the stars for thousands of years. **Stars** are formed when clouds of gas (mainly hydrogen) get compressed. As the hydrogen is compressed, the temperature rises. Eventually, the temperature is so high that the nuclei of hydrogen atoms fuse together, forming helium. This nuclear fusion process emits immense amounts of heat and light. Of course, the Sun is the nearest star to Earth. But there are billions of other stars in the sky. Most of these stars are too far away to see, but if you look at the sky on a clear night, you will see hundreds of them.

People who watch and study the stars are called **astronomers**. Astronomers have grouped bright stars together in patterns and called them **constellations**. These constellations help astronomers and other star watchers to identify and find stars in the sky. Figure 12.2 shows the stars in the constellation Orion. In Greek mythology, Orion was a great hunter.

❶ What is the difference between a star and a planet?

❷ Put the following in order of size from smallest to largest.

   asteroid   galaxy   meteorite   planet   star   Universe

❸ Why can't you see millions of stars during the day?

❹ Why is the idea of constellations helpful?

Activity – Stargazing

The best time to observe the stars is on a dark, cloudless night, about two hours after sunset. It is often difficult to stargaze in cities because of the glare from city lights. It's much better to stargaze in country areas. It will take 15 or 20 minutes for your eyes to adjust to the dark, so don't give up if you can't see much at first.

Wear warm clothes for this activity because you will get cold sitting or standing still. Take a notepad, a pencil and a torch plus binoculars or a telescope. And remember, it's always important to go stargazing with an adult that you know.

❶ Use your binoculars or telescope to find the two well known constellations, Ursa Major and Casseopeia, close to the Pole Star. These constellations can be seen all year round.
First of all, look for Ursa Major, which is often called the 'Plough' or the 'Great Bear' (Figure 12.3). The two stars on the right of Ursa Major are known as pointers. These point towards the Pole Star.
❷ Sketch Ursa Major in your notepad as you see it.
❸ Look carefully at Ursa Major. Which one of the eight stars is, in fact, a double star?
❹ Now, find Casseopeia (Figure 12.4) and draw it.
❺ Make a sketch to show the position of the Pole Star, Ursa Major and Casseopeia.

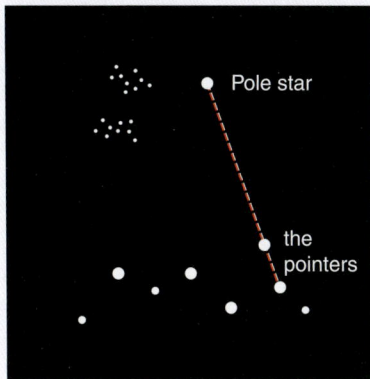

Figure 12.3 The constellation Ursa Major and the Pole Star

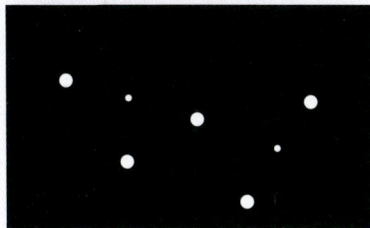

6 If you watch the sky at night over a period of time, the positions of all stars and constellations change, except for the Pole Star. Why is this? (Remember that the Earth spins on an axis through the North and South Poles.)

7 Binoculars and telescopes help you to see much more detail in the night sky. Binoculars and telescopes have their own advantages. Which is better for being:
a) more powerful;   b) easier to use;   c) less expensive;
d) quicker to focus;   e) able to give a steadier image?

8 Now, try to find Orion and other constellations in the night sky. The best way to find constellations is from a sky map (star map). Go to the BBC Sky at Night website (log onto www.bbc.co.uk and search for 'sky at night' then search on 'sky maps'). This will allow you to view a star map for the period when you are looking at the sky. Alternatively, you can go to the same website and click on 'My space' followed by 'constellation guide' to view a particular constellation.

Figure 12.4 The constellation Casseopeia

Careful observation of the stars and constellations using simple telescopes stimulated the early astronomers to make more investigations and this led to new discoveries.

Astronomers found that stars and constellations are not randomly scattered throughout the Universe. They are concentrated in large groups known as **galaxies**. Our Sun is just one of the 100 000 million stars in the Milky Way galaxy (Figure 12.5). If you could view the Milky Way from the side, it would look like two fried eggs stuck back to back – long and thin with a bulge in the middle. But, if you could view the Milky Way from above, it would look like a giant whirlpool with spiral arms. The Milky Way is spinning round slowly and our Sun will take about 220 million years to go round the centre of the galaxy. In millions of years' time, astronomers will see different stars at night.

The distances between stars and between galaxies are so large that it is pointless to measure them in kilometres. The distances are measured in **light years**. One light year is the distance that light travels in one year. The speed of light is 300 000 km/s, so one light year is roughly ten million million (10^{13}) km. The Milky Way is about 100 000 light years across. This means that it takes 100 000 years for light to cross its spirals.

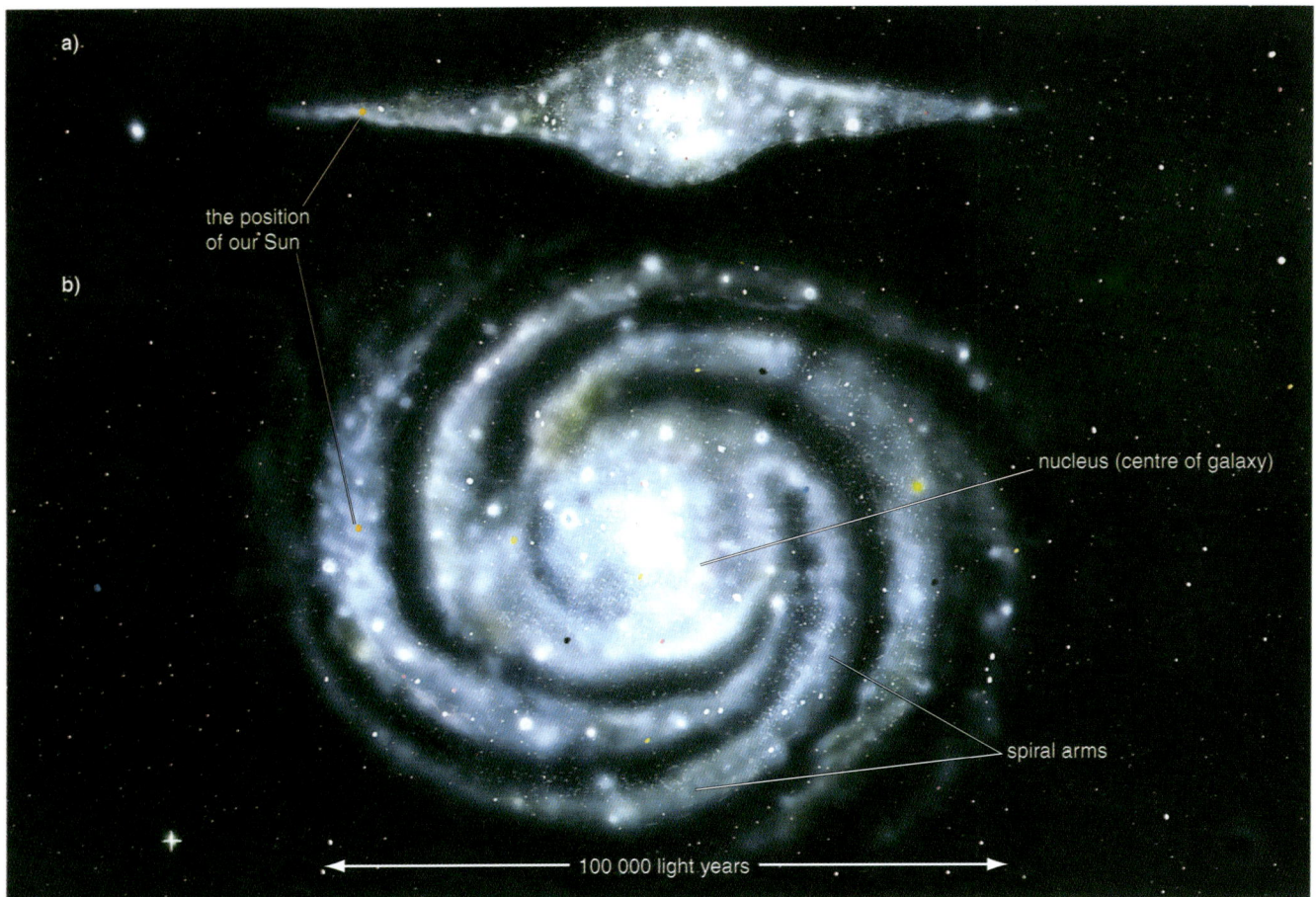

Figure 12.5 a) A side view of the Milky Way and b) a top view of the Milky Way. On a clear, dark night, it is sometimes possible to see a broad, milky band of stars stretching across the sky. This is part of the Milky Way galaxy.

Activity – The Milky Way and other galaxies

Discover more about the Milky Way and other galaxies by going to the Discovery School Website at www.discoveryschool.com.

Then search on 'Galaxy Tour' or 'Milky Way'. As you view the website, plan and write a short report, of about 250 words, on the Milky Way and / or other galaxies.

12.2 Space watch

For thousands of years, people have gazed at the sky and wondered about the mysteries of the Universe. Since the Middle Ages, astronomers have used telescopes and binoculars to view the stars and constellations more clearly and more scientifically.

Today, extremely powerful telescopes allow astronomers to observe and study the Solar System, stars and galaxies throughout the Universe in great detail. The observations are made with telescopes that detect visible light or other electromagnetic radiations, such as radio waves and

X-rays. There are only two different types of powerful telescope – optical telescopes and radio telescopes – but some of these telescopes are based on Earth while others are mounted on artificial satellites orbiting in space.

Optical telescopes

Figure 12.6 The Keck telescopes in Hawaii are two of the largest optical telescopes in the world. Each one is eight storeys tall.

Optical telescopes produce images of stars and galaxies using light. These images are magnified using lenses and mirrors. Very large, powerful optical telescopes are housed in buildings called observatories. The Keck optical telescopes in Hawaii and the Hale optical telescope at Mount Palomar, in the United States, are so powerful that they are never pointed at anything as close as the Moon. These telescopes use huge concave mirrors with diameters of five metres or more. They are so sensitive that they can detect a candle flame 20 000 kilometres away and stars that we cannot even see. Some of these stars are so far away that their light takes millions of years to reach the Earth. When astronomers look at these stars, they are seeing them as they were millions of years ago.

It is very difficult to construct the large, accurate mirror in an optical telescope. Optical telescopes also need huge structures to support the mirror and allow it to move into different positions to pick up light from different parts of the sky.

The main problem with optical telescopes is getting the final image in a position where it can be seen without interference from incoming light rays. This is solved by placing a small flat mirror at an angle just in front of the focus of the concave mirror. The image can then be viewed through an eyepiece (eye lens) away from the incoming light rays (Figure 12.7).

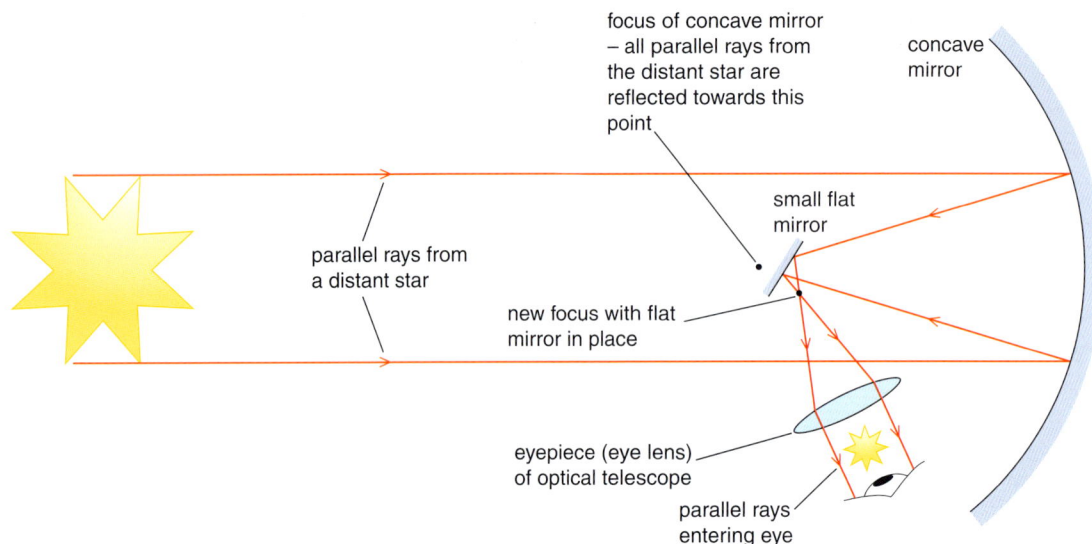

Figure 12.7 The path of light rays in an optical telescope

Radio telescopes

In 1932, an American electrical engineer, called Karl Jansky, detected radio waves coming from space. This simple observation led to the development of radio telescopes. A new area of technology opened up leading to the discovery of pulsars and new galaxies. Pulsars are small stars which rotate quickly, giving off pulses of radiation.

Radio telescopes enable astronomers to study stars and galaxies even further away in space. These distant stars and galaxies give off radio waves, but they cannot be seen using optical telescopes. Radio waves are electromagnetic waves (tiny electrical and magnetic vibrations) which belong to the same family as light but they have much longer wavelengths and lower frequencies (see Section 11.2).

Radio telescopes have a huge concave dish to collect the radio waves given off by certain stars and galaxies (see Figure 12.8). The main dish directs the radio waves towards a secondary reflector, supported above the main dish. This, in turn, reflects them onto a receiver just above the centre of the main dish.

Like many other scientific measurements, there is always some variation in the strength and frequency of radio waves from stars and galaxies.

In order to make their data more reliable, scientists use rows (arrays) of several radio dishes and make repeated observations over days, months and even years.

Signals picked up by several radio telescopes over long periods are often fed into computers. The signals can then be converted into images ('pictures') of the stars and galaxies that are giving off the radio waves.

Figure 12.8 This group of 27 radio telescopes in New Mexico, USA is used to study stars and galaxies beyond the Milky Way. It is known as the Very Large Array.

Space telescopes

Since 1990, scientists have discovered how to put telescopes into space so that they orbit the Earth. These so-called space telescopes can be either optical telescopes or radio telescopes.

Space telescopes can be used to see much further and more clearly than telescopes based on Earth. This is because the Earth's atmosphere does not distort or block their reception of signals from space. By using space telescopes with larger, more carefully constructed mirrors, astronomers can measure smaller and more distant objects in space. Space telescopes are also more sensitive to weaker signals of light, radio waves and other electromagnetic radiations with lower amplitude and lower energy.

The first major space telescope was launched in April 1990. It is called the Hubble Space Telescope, or HST for short (Figure 12.1). It was named in recognition of Edwin Hubble (Section 12.4). HST is an optical telescope which orbits the Earth about 600 km above the ground. Astronomers on Earth can send instructions to cameras on board HST via radio antennae attached to its structure. The same radio antennae relay data and pictures of distant galaxies and planets back to Earth.

At present, scientists in Europe and America are working on the construction of space telescopes that have even more sophisticated functions than the Hubble Space Telescope. The USA's replacement for HST, named The Next Generation Space Telescope, is due for launch in 2009.

In order to understand space exploration further, we must now study what happens to sound and light waves when the source of the waves is moving.

5 What do you understand by the terms
a) optical telescope;
b) radio telescope?

6 What are radio waves?

7 Make a list of the relative advantages and disadvantages of optical telescopes and radio telescopes.

8 a) What is the big advantage of space telescopes over telescopes based on Earth?
b) Suggest three disadvantages of space telescopes.

12.3 What happens to sound waves and light waves when the wave source is moving?

Have you noticed as a police car or a fire engine rushes past you, that the noise of the siren changes? Figure 12.9 will help you to understand why this happens.

In Figure 12.9 a), the source of sound, S, is stationary. Waves spread out from S and someone standing at A hears exactly the same sound as someone at B because they both receive the same number of waves every second.

In Figure 12.9 b), the source is moving to the right. This distorts the waves which become closer together on the right and more spread out on the left. So, someone standing at C hears fewer waves per second than someone at D. The waves at C have a lower frequency and a longer wavelength than those at D.

A similar effect happens with other wave motions, such as light and radio waves, if the wave source is moving very quickly. If a source of light is moving away from you, the wavelength of the light increases. And if the source of light moves even faster, then the wavelength increases even more.

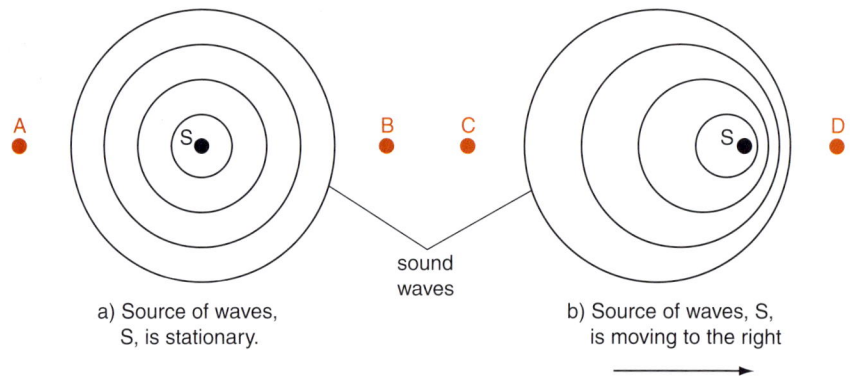

a) Source of waves, S, is stationary.

sound waves

b) Source of waves, S, is moving to the right

Figure 12.9 The sound waves from   a) a stationary source;   b) a moving source

12.4 Moving stars and galaxies

Stargazers have been aware that stars moved across the sky for thousands of years. Early in the twentieth century, astronomers realised that our Solar System was just a very small part of an enormous galaxy of stars. They also realised that the Universe contained millions upon millions of other galaxies.

In 1929, the American astronomer, Edwin Hubble, began to observe and measure these other galaxies. Hubble discovered two very important rules from his observations.

1 The wavelength of light from distant galaxies was *longer* than expected. Hubble called this the **red-shift** because red light has a longer wavelength than other colours and the wavelengths of light from other galaxies moved towards the red end of the spectrum.

2 The further a galaxy is from the Earth, the greater is its red-shift. This is usually called **Hubble's Law**.

12.5

The origins of the Universe

During the middle of the twentieth century, two theories were suggested for the origins of the Universe. These were called the **'big bang' theory** and the **'steady state' theory**.

The 'big bang' theory suggested that the Universe began with an immense explosion billions of years ago. At that moment, all matter that had been concentrated at one small point exploded violently and started to expand. Dust and gases were forced outwards eventually forming the stars and planets that exist today. Since the 'big bang', the Universe has continued to expand, as stars and galaxies move further apart with increasing speed.

The 'steady state' theory suggested that matter was being created continuously in an expanding Universe.

Any theory about the origin of the Universe must account for Hubble's observations. (See Section 12.4.) Hubble explained his first observation by suggesting that the red-shift occurred because distant galaxies were moving away from our own galaxy at great speed. The light emitted from them appears to have a longer wavelength because the waves are being pulled out.

Hubble explained his second observation by suggesting that galaxies further away must be moving faster.

Hubble's red-shift observations provided evidence for a rapidly expanding Universe and supported the 'big bang' theory better than the 'steady state' theory.

Further evidence for the 'big bang' theory came in 1965 with the discovery of cosmic background radiation. The whole Universe appears to emit radiation, which is thought to be a distant echo of the 'big bang'. Today, most astronomers support the 'big bang' theory.

Science and scientists have come a long way, since the early stargazers, in answering questions about the Universe and its origins. They have answered questions about the Sun and its planets, about galaxies and the way they move. But there are limitations to science and lots of questions about the Universe that scientists still cannot answer. The 'big bang' theory is only a theory – a good idea that helps to explain some

❾ How is a source of light moving if its wavelength is longer than expected?

❿ What can you conclude about distant galaxies if the wavelength of light from them is longer than expected?

⓫ Galaxies in the constellation of Ursa Major have a greater red-shift than galaxies in the constellation of Virgo. What can you conclude from this?

Figure 12.10 This galaxy in the constellation of Ursa Major is moving away from us at a speed of 15 000 kilometres per second.

Figure 12.11 The Andromeda Galaxy is our nearest galaxy. Even so, it is more than 2 million light years away from the Earth.

important observations. It may or may not be correct. Science cannot explain why the 'big bang' happened or when the Universe will end.

Activity – Estimating the age of the Universe

Measurements of red-shift allow astronomers to calculate the speed that galaxies are travelling away from us. Table 12.1 shows the speeds of five different galaxies and their distances from Earth.

Galaxy	Speed of galaxy /km per second	Distance of galaxy from Earth/millions of light years
A	6 700	400
B	15 000	900
C	20 000	1 200
D	40 000	2 400
E	60 000	3 600

Table 12.1

❶ What is the range of speeds for the five galaxies?
❷ Is the distance of galaxies from Earth a continuous variable, an ordered variable or a categoric variable?

❸ Table 12.1 does not show clearly any pattern or relationship between the speed of a galaxy and its distance from Earth. Would it be best to draw a bar chart, a line graph or a scattergram to find the relationship between these two variables?
❹ Present the results in Table 12.1 using the method you chose in question 3.
❺ Another galaxy, F, moves away from Earth at 30 000 km/s.
   a) Use the results in Table 12.1 or your answer to question 4 to estimate the distance of galaxy F from Earth.
   b) How long, in millions of years, would it take light to travel this distance?
   c) Galaxy F is, in fact, moving at one tenth ($\frac{1}{10}$) of the speed of light. So, it has taken 18 000 million years for Galaxy F to get to its present distance from Earth. What is the significance of this value, assuming that *all* matter in the Universe was originally in one place?

Summary

✓ **Stars** are huge balls of gas in which the nuclei of hydrogen atoms are fusing together to form helium. This nuclear fusion process emits immense amounts of heat, light and other electromagnetic radiations.

✓ People who watch and study the stars are called **astronomers**.

✓ Careful observations of the stars by early astronomers stimulated further investigations which led to the identification of patterns of bright stars called **constellations**.

✓ Further observation of the Universe with very powerful optical telescopes led to the discovery of **galaxies** (clusters of millions of stars).

✓ The detection of radio waves coming from space led to the development of radio telescopes and the discovery of pulsars and more galaxies.

✓ The distances between stars and galaxies are so large that they are measured in **light years**. One light year is the distance that light travels in one year. This is roughly 10^{13} km.

✓ Space telescopes in orbit above the Earth can identify more distant objects and detect weaker signals than telescopes based on Earth. This is because there is no distortion or interference from the Earth's atmosphere.

✓ The wavelength and frequency of sound waves, light waves and waves from other electromagnetic radiations change when the wave source is moving.

✓ Light and other electromagnetic radiations from distant galaxies have a longer wavelength than expected. This is known as the **red-shift**.

✓ Galaxies further away from the Earth give a greater red-shift than those closer to the Earth.

EXAMQUESTIONS

❶ Match the following questions with answers A, B, C and D below.
   a) Which bodies emit light?
   b) Which bodies are satellites of stars?
   c) Which bodies are satellites of planets?
   d) Which bodies are smaller than moons?
   e) Which bodies provide evidence for the origin of the Universe?
   A Galaxies   **B** Meteorites
   C Moons   **D** Planets   (*5 marks*)

❷ Which of the following statements about the 'big bang' theory are true and which are false?
   A It has been proved correct by mathematical calculations.
   B It is supported by the fact that distant galaxies are moving away from the Earth.
   C It is based on scientific and religious facts.
   D It is the most satisfactory explanation of present scientific knowledge.
   E It is the only way to explain the origin of the Universe.   (*5 marks*)

❸ a) Radio telescopes are used to investigate distant galaxies. Why are radio telescopes much larger than optical telescopes?   (*2 marks*)
   b) Why are space telescopes usually more sensitive than earthbound telescopes?   (*1 mark*)
   c) The light from distant galaxies shows a red-shift.
      i) Explain the term 'red-shift'.   (*3 marks*)
      ii) What does the red-shift effect tell us about distant galaxies?   (*1 mark*)
      iii) Galaxies further away from the Earth show a larger red shift than galaxies closer to the Earth. What does this tell us about galaxies?   (*1 mark*)

❹ The 'big bang' is one theory about the origin of the Universe.
   a) State the main ideas of the 'big bang' theory.   (*3 marks*)
   b) How do Hubble's observations of the red-shift of distant galaxies support the 'big bang' theory?   (*1 mark*)
   c) How does background radiation from the whole Universe support the 'big bang' theory?   (*1 mark*)
   d) Why are some astronomers unsure about the 'big bang' theory?   (*2 marks*)

❺ Figure 12.12 shows a side view of two thin beams of radio waves approaching the dish of a radio telescope.
   a) Redraw the figure showing the path of the two thin beams until they reach the receiver.   (*2 marks*)
   b) Why is it necessary to use a secondary reflector?   (*1 mark*)

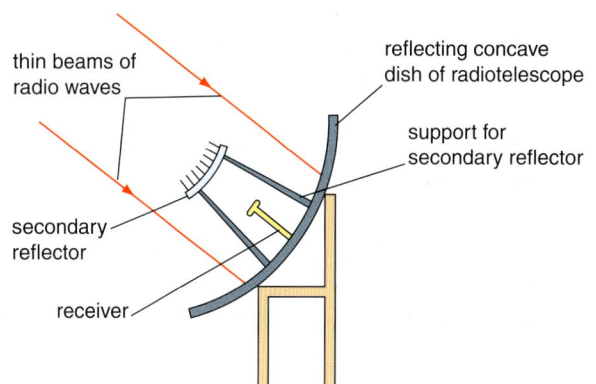

Figure 12.12

Index

absorbers (radiation) 161–3
accuracy 191
acid rain 112, 113, 185–6
acidity 88
adaptation 39–40
addictive drugs 13, 15
additives (food) 131–3
ADH *see* anti-diuretic hormone
adrenaline 6
air pollution 112–14
alcohol 15, 118
algae 44
alkanes 106–8
alkenes 116–17
alloys 79, 96–9
alpha particles 214–16
aluminium 97
amino acids 19, 20
amplitude (waves) 199–200
anaerobic bacteria 47
analogue signals 210–11
anomalous results 163
anorexia 30
antibiotics 34
antibodies 33
anti-diuretic hormone (ADH) 7
antigens 33
Arctic 39
arid regions 39
arthritis 30
asexual reproduction 64
astronomers 224–5
atmosphere 140–58
atomic numbers 81, 83–4, 212–13
atoms 80–3, 211–13

background radiation 216
bacteria 32, 34, 64, 67
balanced equations 85
Becquerel, Henri 214
beta particles 214–16
'big bang' theory 231–2
biodegradable materials 43–4
biodiesel 137, 138
biofuels 186
bioindicators 53–4, 56
biological control species 42
bird flu 35
blast furnaces 94–5
blending theory 61
blood sugar 7
BMI *see* body mass index
bodies
   heat transfer rates 166–8
   protection (pathogens) 33–6
   responses to change 1–17
body mass index (BMI) 29, 31

bonds 81
brown field sites 44, 45
Buffon, George Leclerc 69–70
building materials 88–91
butane 106–7
butter 133

calcium 19
camels 39
cannabis 13
carbon dioxide 47, 152–3, 155–6, 185
catalytic cracking 114–15
catastrophism 70
categoric variables 160
cattle ranching 47
cauliflowers 68
cement 90
central nervous system (CNS) 2–5
characteristics 60–2
chemical reactions 78
chemical symbols 80, 84–7
chemical transmitters 4
cholesterol 22
CHP *see* combined heat and power
chromosomes 63
chromatography 133
climate change 112–14
clinical trials 12
cloning 64
close packing (atoms) 98
CNS *see* central nervous system
cocaine 15
cold 165–6
combined heat and power (CHP) 175
communications 207–9, 210–11
competition 40–1
composting 43
compounds 79–82
concrete 90
conducting polymers 125
conduction 164–6
constellations 224
contact lenses 124
continental drift 145
continuous variables 160
contraception 9
control variables 20
convection 164–6
cooking 202–3
copper 97
core (Earth) 141–3
cracking (crude oil) 114–15
creationism 69
crude oil 103–27
crust (Earth) 141–2, 149
Curie, Marie 214
cuttings 64

Cuvier, George 70

dams 189
Darwin, Charles 70–1
decomposition 88–9
deforestation 45–7
degradable plastics 122
dentistry 124
dependent variables 162–3
diabetes 29
diets 31
digital signals 210–11
discrete variables 160
disease 31–2
distillation 104–5
dodo 73–4
dominant characteristics 62
double bonds 108, 134
drugs 11–15
dwarfism 67

E-numbers 131
Earth 79, 140–58, 195
earthquakes 148, 149, 150–2
Easter Island 52
ecosystems 42, 47
EFAs *see* essential fatty acids
effectors 2
efficiency 159–77
EfW (Energy from Waste) 43
electricity 3–4, 178–97
electrolysis 93, 100
electromagnetic waves 198–222
electrons 81–2, 211–12
elements 78–84
embryo transplant 65
emitters (radiation) 160–1
emulsions/emulsifiers 129–30
energy
   balance 27
   conduction/convection 164–6
   conservation of 171
   from Earth 195
   efficiency 50, 159–77
   electricity 178–97
   food 20
   National Grid 182–3
   non-renewable 183–6
   radiation 160–3
   renewable 183, 187–94
   from Sun 194
   transfer 160–6, 171–5
   transformation 171–5
energy-efficient lamps 204–5
energy-saving ideas 169–70
environmental factors 27
equations 78, 85